Gene Ontology: A Resource for Analysis and Interpretation of Biological Database

Gene Ontology: A Resource for Analysis and Interpretation of Biological Database

Edited by Paige Hill

New York

Published by Syrawood Publishing House,
750 Third Avenue, 9th Floor,
New York, NY 10017, USA
www.syrawoodpublishinghouse.com

Gene Ontology: A Resource for Analysis and Interpretation of Biological Database
Edited by Paige Hill

International Standard Book Number: 978-1-64740-380-5 (Hardback)

Cataloging-in-publication Data

Gene ontology : a resource for analysis and interpretation of biological database / edited by Paige Hill.
p. cm.
Includes bibliographical references and index.
ISBN 978-1-64740-380-5
1. Genes. 2. Molecular genetics. 3. Bioinformatics. 4. Molecular biology. I. Hill, Paige.
QH447 .G46 2023
572.86--dc23

TABLE OF CONTENTS

PREFACE

The gene ontology (GO) is a bioinformatics project which aims to merge the representation of gene and characteristics of gene product in all species. It is a thorough and updated compilation of biological information that is utilized as a resource for computational representation of the most recent scientific data available concerning coding as well as non-coding functions of genes. High-throughput biological datasets can be interpreted and analyzed using the GO resource. The goal of the GO analysis is to pinpoint the molecular functions, biological processes and cellular locations of the genes. It utilizes markup language in order to make the genetic data and curated attributes of gene products readable by machine, in a manner which is integrated across all species. This book provides comprehensive insights on gene ontology. It aims to shed light on some of the unexplored aspects of this subject. Those in search of information to further their knowledge will be greatly assisted by this book.

Significant researches are present in this book. Intensive efforts have been employed by authors to make this book an outstanding discourse. This book contains the enlightening chapters which have been written on the basis of significant researches done by the experts.

Finally, I would also like to thank all the members involved in this book for being a team and meeting all the deadlines for the submission of their respective works. I would also like to thank my friends and family for being supportive in my efforts.

Editor

PREFACE

Fundamentals

1

The Gene Ontology and the Meaning of Biological Function

Paul D. Thomas

Abstract

The Gene Ontology (GO) provides a framework and set of concepts for describing the functions of gene products from all organisms. It is specifically designed for supporting the computational representation of biological systems. A GO annotation is an association between a specific gene product and a GO concept, together making a statement pertinent to the function of that gene. However, the meaning of the term "function" is not as straightforward as it might seem, and has been discussed at length in both philosophical and biological circles. Here, I first review these discussions. I then present an explicit formulation of the biological model that underlies the GO and annotations, and discuss how this model relates to the broader debates on the meaning of biological function.

Key words Genome, Function, Ontology, Selected effects, Causal role

1 What Is Biological Function?

The notion of function in biology has received a great deal of attention in the philosophical literature. At the broadest level, there are two schools of thought on how functions should be defined, now most commonly referred to as "causal role function" and "selected effect function." Causal role function was first proposed by Cummins [1], and it focuses on describing function in terms of how a part contributes to some overall capacity of the system that contains the part. In this formulation, the function of an entity is relative to some system to which it contributes. For example, the statement "the function of the heart is to pump blood" has meaning only in the context of the larger circulatory system's capacity to deliver nutrients and remove waste products from bodily tissues. However, one of the main objections to the causal role definition of function is that there is no systematic way to identify what the larger system (and the relevant capacity of that system) should be. Selected effect function, on the other hand, derives from the "etiological" definition of function first proposed by Wright [2]. In this formulation, a function of an entity is the ultimate answer to the question of why the entity exists at all. In biology, as explained by Millikan [3] and Neander [4], this is tantamount to asking the following: For which of its effects was it selected during evolution? One obvious advantage of the selected effect definition is that it explicitly incorporates evolutionary considerations, and demands that a function ultimately derive from its history of natural selection. On the more practical side, it has the further advantage of putting constraints on which effects, out of the myriad causal effects that a particular entity might have, could be considered as functions. Following the example above, an effect of the heart (beating) is to produce a sound, but it would not be correct to say that the function of the heart is to produce a sound. The selected effects definition of function would distinguish a proper function (e.g., pumping blood) from an "accidental" effect (e.g., producing a sound) on the basis that natural selection more likely operated on the heart's effect of pumping blood. In the causal role definition, on the other hand, there is always the potential for arbitrariness and idiosyncrasy in defining a containing system and capacities; thus there is no general rule for distinguishing functional from accidental effects.

Nevertheless, causal role function has been stalwartly defended by biologists in the subdiscipline of functional anatomy [5], which emphasizes how anatomical parts function as parts of larger systems. They claim that the selected trait can be difficult to infer, and lack of a hypothesis for such a trait should not stand in the way of an analysis of the mechanism of how an anatomical feature operates. For example, one could analyze a jaw in terms of its capacity for generating a crushing force irrespective of whether it was selected for crushing seeds or defending against a predator. Indeed, the search for mechanisms of operation, or more generally just "mechanism," has more recently been offered as an alternative paradigm for molecular and neurobiology in particular [6]. Mechanism, like causal role, focuses on how parts contribute to a system. But it takes a step further in defining core concepts, and how these relate to function. The core concepts are entities and activities: physical entities (such as proteins) perform activities, or actions that can have causal effects on other activities. In this view, a function is simply an activity that is carried out as part of a larger mechanism. For example, the function of the ribosome (an entity) is translation (an activity), and translation plays a role in a larger mechanism of gene expression. The subtle difference from earlier formulations of function is an emphasis on *the activity having the role of a function*, rather than *the entity itself having a function*. Also like causal role, no a priori constraints are put on mechanism: "a function is … a component in some mechanism, that is … in a context that is taken to be important, vital, or otherwise significant." Clearly mechanism is susceptible to the same criticism as causal role function, regarding arbitrariness in the choice of system.

The core differences between selected effect function and causal role function derive largely from differences in what question they are trying to answer. For selected effect function, the question is about origins: Why is the entity there (i.e., what explains its selective advantage)? [2]. For causal role function, the question is about operation: How does the entity contribute to the biological capacities of the organism that has the entity (and only secondarily, how do those capacities relate to natural selection)? [1]. And there is little doubt that in most biological research endeavors today, the concern is in elucidating the mechanisms by which biological systems operate, rather than in explaining why the parts are there to begin with.

The notion of function, particularly in connection with molecular biology, has been discussed at length not only by philosophers, but also by molecular biologists themselves. As a representative sample, I will consider two publications written with very different aims in mind: a textbook chapter by Alberts entitled "Protein Function" [7] and a philosophical treatise by Monod, *Chance and Necessity* [8]. Alberts' treatment of "function" covers two distinct but related senses of the word. The first is how an individual protein *works* at the mechanistic level (its manner of functioning): "how proteins bind to other selected molecules and how their activity depends on such binding." The second is to describe how a protein acts as a component in a larger system, by analogy to mechanical parts in human-designed systems (its functional role in the context of the operation of the cell): "proteins ... act as catalysts, signal receptors, switches, motors, or tiny pumps." Specific molecular binding can be considered the general mechanism by which a functional role can be carried out. These uses of "function" appear, at least on the face of it, to be more in line with the causal role and mechanism views in the philosophical literature.

Given its broader intended audience of scientists and laymen (and presumably philosophers), *Chance and Necessity* puts biological function in a much broader context. Monod coins the term "teleonomic function" to describe more precisely what he means by function. He carefully defines teleonomy as the characteristic of "objects endowed with a purpose or project, which at the same time they exhibit through their structure and carry out through their performances" [p. 9]. Teleonomy is also a property of human-designed "artifacts," further emphasizing the view of function in terms of an apparent purpose in accomplishing a predetermined aim. But living systems owe their teleonomy to a distinct source. As he so eloquently (if also compactly) states, "invariance necessarily precedes teleonomy" [p. 23], which he goes on to explain further as "the Darwinian idea that the initial appearance, evolution and steady refinement of ever more intensely teleonomic structures are due to perturbations in a structure *which already possesses the property of invariance.*" Thus what appears to be a future-goal-oriented action by a living organism is, in fact, only a blind repetition of a genetic program that evolved in the past. Importantly, Monod notes the presence of teleonomy at all levels of a biological system, from proteins (which he calls "the essential molecular agents of teleonomic performance") to "systems providing large scale coordination of the organism's performances ... [such as] the endocrine and nervous systems" [p. 62]. In this way, Monod's teleonomic function includes aspects of both Wright's selected effect function (the origin of apparently designed functions in prior natural selection) and Cummins's causal role function (the role of a part in a larger system).

In summary, function as conceived by molecular biologists (in what could be called the "molecular biology paradigm") refers to specific, coordinated activities that have the appearance of having been designed for a purpose. That apparent purpose is their function. The appearance of design derives from natural selection, so many biologists now favor the use of the term "biological program" to avoid connotations of intentional design. Following this convention, biological programs, when executed, perform a function; that is, they result in a particular, previously selected outcome or causal effect. Biological programs are nested modularly inside other, larger biological programs, so a protein can be said to have functions at multiple levels. The lowest level biological program is expression of a single macromolecule, e.g., a protein: the gene is transcribed into RNA, which is translated into a protein, which adopts a particular structure that performs its function simply by following physical laws that determine how it will interact with specific (i.e., a small number) of other distinct types of other molecular entities. At higher levels, the functions of multiple proteins are executed in a coherent, controlled ("regulated") manner to accomplish a larger function. Thus, simply identifying a coherent, regulated system of activities can be a fruitful, practical start for identifying selected effect functions. Causal role analyses can and do play such a role in functional anatomy and molecular biology. But of course they are only *candidates* for evolved biological functions until they have been related to past survival and reproduction, the ultimate function of every biological program.

2 Function in the Gene Ontology

I now turn to a description of how function is conceived of, and represented in practice, in the Gene Ontology.

2.1 Gene Products, Not Genes, Have Functions

In order to understand how gene function is represented in the GO, some basic molecular biology knowledge is required.

- A *gene* is a contiguous region of DNA that encodes instructions for how the cell can make a large ("macro") molecule (or potentially multiple different macromolecules).

- A macromolecule is called a *gene product* (as it is produced deterministically according to the instructions from a gene), and can be of two types, a *protein* (the most common type) or a *noncoding RNA*.
- A gene product can act as a molecular machine; that is, it can perform a chemical action that we call an *activity*.
- Gene products from different genes can combine into a larger molecular machine, called a macromolecular *complex*.

Each concept in the Gene Ontology relates to the activity of a gene product or complex, as these are the entities that carry out cellular processes. A gene encodes a gene product, so it can obviously be considered the ultimate source of these activities and processes. But strictly speaking, a gene does not perform an activity itself. Thus, when the Gene Ontology refers to "gene function," it is actually shorthand for "gene product function."

2.2 Assertions About Functions of Particular Genes Are Made by "GO Annotations"

The Gene Ontology defines the "universe" of possible functions a gene might have, but it makes no claims about the function of any particular gene. Those claims are, instead, captured as "GO annotations." A GO annotation is a statement about the function of a particular gene. But our biological knowledge is extremely incomplete. Accordingly, the GO annotation format is designed to capture partial, incomplete statements about gene function. A GO annotation typically associates only a single GO concept with a single gene. Together, these statements comprise a "snapshot" of current biological knowledge. Different pieces of knowledge regarding gene function may be established to different degrees, which is why each GO annotation always refers to the evidence upon which it is based.

2.3 The Model of Gene Function Underlying the GO

The Gene Ontology (GO) considers three distinct aspects of how gene functions can be described: **molecular function, cellular component**, and **biological process** (note that throughout this chapter, **bold text** will denote specific concepts, or classes, from the Gene Ontology). In order to understand what these aspects mean and how they relate to each other, it may be helpful to consider the biological model assumed in GO annotations. GO follows what could be called the "molecular biology paradigm," as described in the previous section. In this representation, a gene encodes a gene product, and that gene product carries out a molecular-level process or activity (**molecular function**) in a specific location relative to the cell (**cellular component**), and this molecular process contributes to a larger biological objective (**biological process**) comprised of multiple molecular-level processes. An example, elaborating on the example in the original GO paper [9], is shown in Fig. 1.

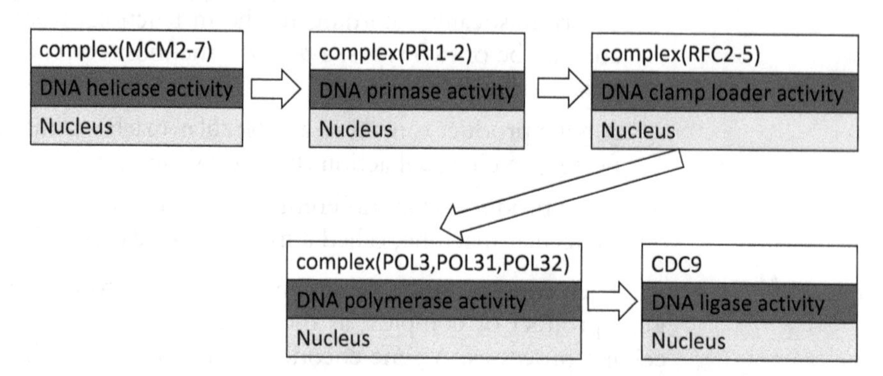

Fig. 1 DNA replication (in yeast) as modeled using the GO. Gene products/complexes (*white*) perform molecular processes (**molecular function**, *red*) in specific locations (**cellular component**, *yellow*), as part of larger biological objectives (**biological process**, specifically **DNA-directed DNA replication**)

To reiterate, GO concepts were designed to apply specifically to the actions of gene products, i.e., *macromolecular machines* comprising proteins, RNAs, and stable complexes thereof. In the GO representation, a region of DNA (e.g., a regulatory region) is treated not as carrying out a molecular process, but rather as an object that gene products can act upon in order to perform their specific activities.

2.4 Molecular Functions Define Molecular Processes (Activities)

In the GO, a **molecular function** is a process that can be carried out by the action of a single macromolecular machine, via direct physical interactions with other molecular entities. Function in this sense denotes an action, or activity, that a gene product performs. These actions are described from the two distinct but related perspectives commonly employed by biologists: (1) biochemical activity, and (2) role as a component in a larger system/process. Biochemical activities include binding and catalytic activities, and are only functions in the broad sense, i.e., how something functions, the molecular mechanism of operation. Component role descriptions, on the other hand, refer to roles in larger processes, and are sometimes described by analogy to a mechanical or electrical system. For example, biologists may refer to a protein that functions (acts) as a **receptor**. This is because the activity is interpreted as receiving a signal, and converting that signal into another physicochemical form. Unlike biochemical activities, these roles require some degree of *interpretation* that includes knowledge of the larger system context in which the gene product acts.

2.5 Cellular Components Define Places Where Molecular Processes Occur

A **cellular component** is a location, relative to cellular compartments and structures, occupied by a macromolecular machine when it carries out a **molecular function**. There are two ways in which biologists describe locations of gene products: (1) relative to cellular structures (e.g., **cytoplasmic side of plasma membrane**) or compartments (e.g., **mitochondrion**), and (2) the stable macromolecular complexes of which they are parts (e.g., the **ribosome**). Unlike the other aspects of GO, **cellular component** concepts refer not to processes but rather a cellular anatomy. Nevertheless, they are designed to be applied to the actions of gene products and complexes: a GO annotation to a **cellular component** provides information about where a molecular process may occur during a larger process.

2.6 Biological Processes Define Biological Programs Comprised of Regulated Molecular Processes

In the GO, a **biological process** represents a specific objective that the organism is genetically "programmed" to achieve. Each **biological process** is often described by its outcome or ending state, e.g., the biological process of **cell division** results in the creation of two daughter cells (a divided cell) from a single parent cell. A **biological process** is accomplished by a particular set of molecular processes carried out by specific gene products, often in a highly regulated manner and in a particular temporal sequence.

An annotation of a particular gene product to a GO **biological process** concept should therefore have a clear interpretation: the gene product carries out a molecular process that plays an integral role in that biological program. But a gene product can affect a biological objective even if it does not act strictly within the process, and in these cases a GO annotation aims to specify that relationship insofar as it is known. First, a gene product can control when and where the program is executed; that is, it might *regulate* the program. In this case, the gene product acts outside of the program, and controls (directly or indirectly) the activity of one or more gene products that act within the program. Second, the gene product might act in another, separate biological program that is *required for* the given program to occur. For instance, animal embryogenesis requires translation, though translation would not generally be considered to be part of the embryogenesis program. Thus, currently a given **biological process** annotation could have any of these three meanings (namely a gene activity could be part of, regulate, or be upstream of but still necessary for, a biological process). The GO Consortium is currently exploring ways to computationally represent these different meanings so they can be distinguished.

Biological process is the largest of the three ontology aspects in the GO, and also the most diverse. This reflects the multiplicity of levels of biological organization at which genetically encoded programs can be identified. **Biological process** concepts span the entire range of how biologists characterize biological systems. They can be as simple as a generic enzymatic process, e.g., **protein phosphorylation**, to molecular pathways such as **glycolysis** or the **canonical Wnt signaling pathway**, to complex programs like **embryo development** or **learning**, and even including **reproduction**, the ultimate function of every evolutionarily retained gene.

Because of this diversity, in practice not all **biological process** classes actually represent coherent, regulated biological programs.

In particular, GO **biological process** also includes molecular-level processes that cannot always be distinguished from molecular functions. Taking the previous example, the process class **protein phosphorylation** overlaps in meaning with the molecular activity class **protein kinase activity**, as protein kinase activity is the enzymatic activity by which protein phosphorylation occurs. The main difference is that while a **molecular function** annotation has a precise semantics (e.g., the gene carries out protein kinase activity), the **biological process** annotation does not (e.g., the gene either carries out, regulates, or is upstream of but necessary for a particular protein kinase activity).

3 How Does the GO Relate to the Debate About the Meaning of Biological Function?

GO concepts are designed to describe aspects (molecular activity, location of the activity, and larger biological programs) of the *functions that a gene evolved to perform*, i.e., selected effect functions. However, GO concepts may not always be applied that way. As a result, a given GO annotation may or may not be a statement about selected effect function. Note that while all biological programs are carried out by molecular activities, not all molecular activities necessarily contribute to a biological program. In principle, then, only those GO annotations that refer to biological programs can be considered to generally reflect selected effect functions.

A GO **molecular function** annotation by itself cannot be automatically interpreted as selected effect function. One of the most vigorous long-standing debates in the GO Consortium concerns the **protein binding** class in GO, as it is clearly appreciated by biologists that a given experimental observation of molecular binding may reflect biological noise and not necessarily contribution to a biological objective. Even further removed, **cellular component** annotations are often made from observations of a protein in a particular compartment, irrespective of whether the protein performs a molecular activity in that location. For example, many proteins known to act extracellularly are also observed in the Golgi apparatus as they await trafficking to the plasma membrane. In short, if the molecular activity and cellular location are not yet implicated in a biological program (that is itself clearly related to survival and reproduction), they cannot be said to have selected effect function. Strictly speaking, such annotations should be considered as referring to *candidate* functions, rather than *proper* functions.

Despite these theoretical considerations, most GO annotations are likely in practice to refer to selected effect functions. This is simply because most GO annotations are made from publications describing specific, small-scale molecular biology studies that focus on a particular biological program. In such studies, a biological objective (usually implicitly related to survival and reproduction) has already been established in advance, and the paper describes the mechanistic activities of gene products in accomplishing that biological objective. Large-scale studies, on the other hand, that measure gene product activities or locations without reference to the biological program they are part of, should be considered as *candidate* selected effect functions. This view would address the recent debate about gene function [10–12], initiated when the ENCODE (Encyclopedia of DNA Elements) project—a large-scale, hypothesis-free project to catalog biochemical activities across numerous regions of the human genome [13]—inappropriately claimed to have discovered proper functions. The GO Consortium is discussing ways to help users distinguish between hypothesis-driven annotations (likely proper functions) from large-scale annotations (candidate functions).

4 Conclusion

It has not generally been appreciated that the Gene Ontology concepts for describing aspects of gene function assume a specific model of how gene products act to achieve biological objectives. My aim here has been to describe this model, which, I hope, will clarify how GO annotations should be properly used and interpreted, as well as how the GO relates to biological function as discussed in both the philosophical and biological literature.

Acknowledgments

I want to thank Christophe Dessimoz, Pascale Gaudet, Jenna Hastings, and Ridhima Kad for helpful discussions on this topic, and for thoughtful reading and suggestions on the manuscript. Open Access charges were funded by the University College London Library, the Swiss Institute of Bioinformatics, the Agassiz Foundation, and the Foundation for the University of Lausanne.

References

1. Cummins R (1975) Functional analysis. J Philos 72:741–765
2. Wright L (1973) Functions. Philos Rev 82: 139–168
3. Millikan RG (1989) In defense of proper functions. Philos Sci 56:288–302
4. Neander K (1991) The teleological notion of "function.". Australas J Philos 69:454–468
5. Amundson R, Lauder GV (1994) Function without purpose: the uses of causal role function in evolutionary biology. In: Hull DL, Ruse M (eds) Biology & philosophy, vol 9. Oxford University Press, Oxford, pp 443–469
6. Machamer P, Darden L, Craver CF (2000) Thinking about mechanisms. Philos Sci 67:1–25
7. Alberts B (2002) Protein function. In: Molecular biology of the cell, 4th edn. Garland Science, New York
8. Monod J (1971) Chance and necessity. Alfred Knopf, New York
9. Ashburner MA et al (2000) Gene ontology: tool for the unification of biology. Nat Genet 25:25–29
10. Doolittle WF (2013) Is junk DNA bunk? A critique of ENCODE. Proc Natl Acad Sci U S A 110:5294–5300
11. Doolittle WF et al (2014) Distinguishing between "function" and "effect" in genome biology. Genome Biol Evol 6:1234–1237
12. Graur D et al (2013) On the immortality of television sets: "function" in the human genome according to the evolution-free gospel of ENCODE. Genome Biol Evol 5:578–590
13. Dunham I et al (2012) An integrated encyclopedia of DNA elements in the human genome. Nature 489:57–74

2

Primer on the Gene Ontology

Pascale Gaudet, Nives Škunca, James C. Hu, and Christophe Dessimoz

Abstract

The Gene Ontology (GO) project is the largest resource for cataloguing gene function. The combination of solid conceptual underpinnings and a practical set of features have made the GO a widely adopted resource in the research community and an essential resource for data analysis. In this chapter, we provide a concise primer for all users of the GO. We briefly introduce the structure of the ontology and explain how to interpret annotations associated with the GO.

Key words Gene Ontology structure, Evidence codes, Annotations, Gene association file (GAF), GO files, Function, Vocabulary, Annotation evidence

1 Introduction

The key motivation behind the Gene Ontology (GO) was the observation that similar genes often have conserved functions in different organisms [1]. Clearly, a common vocabulary was needed to be able to compare the roles of orthologous genes (and their products) across different species. The value of comparative studies of biological function across systems predates Jacques Monod's statement that "anything found to be true of *E. coli* must also be true of elephants" [2]. The Gene Ontology aims to produce a rigorous shared vocabulary to describe the roles of genes across different organisms [1]. The GO project consists of the *Gene Ontology* itself, which models biological aspects in a structured way, and *annotations*, which associate genes or gene products with terms from the Gene Ontology. Combining information from all organisms in one central repository makes it possible to integrate knowledge from different databases, to infer the functionality of newly discovered genes, and to gain insight into the conservation and divergence of biological subsystems.

In this primer, we review the fundamentals of the GO project. The chapter is organised as answers to five essential questions: What is the GO? Why use it? Who develops it and provides annotations? What are the elements of a GO annotation? And finally, how can the reader learn more about GO resources?

2 What Is the Gene Ontology?

The Gene Ontology is a controlled vocabulary of terms to represent biology in a structured way. The terms are subdivided into three distinct ontologies that represent different biological aspects: Molecular Function (MF), Biological Process (BP), and Cellular Component (CC) [1]. These ontologies are non-redundant and share a common space of identifiers and a well-specified syntax.

Terms are linked to each other by relations to form a hierarchical vocabulary (Chap. 1 [3]). This is often modelled as a graph in which the relationships form the directed edges, and the terms are the nodes (Fig. 1). Since each term can have multiple relationships to broader parent terms and to more specific child terms, the structure allows for more expressivity than a simple hierarchy.

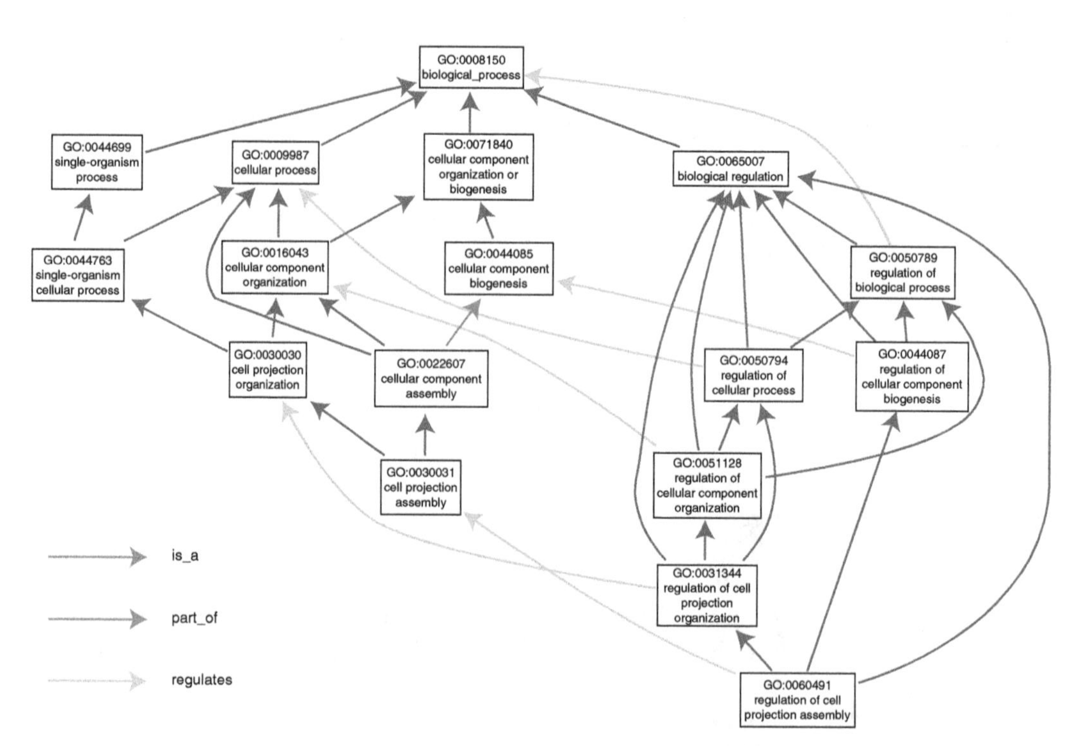

Fig. 1 The structure of the Gene Ontology (GO) is illustrated on a subset of the paths of the term "regulation of cell projection assembly", GO:0060491, to its root term. The GO is a directed graph with terms as nodes and relationships as edges; these relationships are either is_a, part_of, has_part, or regulates. In its basic representation, there should be no cycles in this graph, and we can therefore establish parent (more general) and child (more specific) terms (Chap. 11 [4] for more details on the different representations). Note that it is possible for a term to have multiple parents. This figure is based on the visualisation available from the AmiGO browser, generated on November 6, 2015 [5]

The full GO is large: in October 2015, the full ontology specification had 43835 terms, 73776 explicitly encoded is_a relationships, 7436 explicitly encoded part_of relationships, and 8263 explicitly encoded regulates, negatively_regulates, or positively_regulates relationships. This level of detail is not necessary for all applications. Many research groups who do GO annotations for specific projects use the generic GO-slim file, which is a manually curated subset of the Gene Ontology containing general, high-level terms across all biological aspects. There are several GO slims,[1] ranging from the general Generic GO slim developed by the GO Consortium to more specific ones, such as the Chembl Drug Target slim.[2]

To keep up with the current state of knowledge, as well as to correct inaccuracies, the GO undergoes frequent revisions: changes of relationships between terms, addition of new terms, or term removal (obsoletion). Terms are never deleted from the ontology, but their status changes to obsolete and all relationships to the term are removed [6]. Furthermore, the name itself is preceded by the word "obsolete" and the rationale for the obsoletion is typically found in the Comment field of the term. An example of an obsolete term is GO:0000005, "obsolete ribosomal chaperone activity". This MF GO term was made obsolete "because it refers to a class of gene products and a biological process rather than a molecular function".[3] Changes to the *relationships* do not impact annotations, because annotations are associated with a given GO term regardless of its relationships to other terms within the GO. Obsoletion of terms however has an impact on *annotations* associated with them: in some cases, the old term can be automatically replaced by a new or a parent one; in others, the change is so important that the annotations must be manually reviewed.

However, these changes can affect the analyses done using the ontology. In articles or reports, it is good practice to provide the version of the file used for a particular analysis. In GO, the version number is the date the file was obtained from the GO site (GO files are updated daily).

3 Why Use the Gene Ontology?

Because it provides a standardised vocabulary for describing gene and gene product functions and locations, the GO can be used to query a database in search of genes' function or location within the cell or to search for genes that share characteristics [7]. The hierarchical structure of the GO allows to compare proteins annotated to different terms in the ontology, as long as the terms have

[1] http://geneontology.org/page/go-slim-and-subset-guide
[2] http://wwwdev.ebi.ac.uk/chembl/target/browser
[3] https://www.ebi.ac.uk/QuickGO/GTerm?id=GO:0000005

relationships to each other. Terms located close together in the ontology graph (i.e. with a few intermediate terms between them) tend to be semantically more similar than those further apart (*see* Chap. 12 on comparing terms [8]).

The GO is frequently used to analyse the results of high-throughput experiments. One common use is to infer commonalities in the location or function of genes that are over- or under-expressed [6, 9, 10]. In functional profiling, the GO is used to determine which processes are different between sets of genes. This is done by using a likelihood-ratio test to determine if GO terms are represented differently between the two gene sets [6].

Additionally, the GO can be used to infer the function of unannotated genes. Gene predictions with significant similarity to annotated genes can be assigned one or several of the functions of the characterised genes. Other methods such as the presence of specific protein domains can also be used to assign GO terms [11, 12]. This is discussed in Chap. 5 [13].

A wealth of tools—web-based services, stand-alone software, and programing interfaces—has been developed for applying the GO to various tasks. Some of these are presented in Chap. 11 [4].

While Gene Ontology resources facilitate powerful inferences and analyses, researchers using the GO should familiarise themselves with the structure of the ontology and also with the methods and assumptions behind the tools they use to ensure that their results are valid. Common pitfalls and remedies are detailed in Chap. 14 [14].

4 Who Develops the GO and Produces Annotations?

The GO Consortium consists of a number of large databases working together to define standardised ontologies and provide annotations to the GO [15]. The groups that constitute the GO consortium include UniProt [16], Mouse Genome Informatics [17], *Saccharomyces* Genome Database [18], Wormbase [19], Flybase [20], dictyBase [21], and TAIR [22]. In addition, several other groups contribute annotations, such as EcoCyc [23] and the Functional Gene Annotation group at University College London [24].[4] Within each group, biocurators assign annotations according to their expertise [25]. Further, the GO Consortium has mechanisms by which members of the broader community (*see* Chap. 7 [26]) can suggest improvements to the ontology and annotations.

[4] Full list at http://geneontology.org/page/go-consortium-contributors-list

5 What Are the Elements of a GO Annotation?

This section describes the different elements composing an annotation and some important considerations about each of them. The annotation process from a curator standpoint is discussed in detail in Chap. 4 [27].

Fundamentally, a GO annotation is the association of a gene product with a GO term. From its inception, the GO Consortium has recognised the importance of providing supporting information alongside this association. For instance, annotations always include information about the evidence supporting the annotation.

Over time, the GO Consortium standards for storing annotations have evolved to improve this representation. Annotations are now stored in one of the two formats: GAF (Gene Association File), and the more recent GPAD (Gene Product Association Data). The two formats contain the same information but there are differences in how the data is normalised and represented (discussed in more details in Chap. 11 [4]). In this primer, we focus on the former. The representation of an annotation in the GAF file format 2.1 is shown in Fig. 2. It contains 17 fields (also sometimes referred to as "columns"). We describe them in this section.

5.1 Annotation Object

The annotation object is the entity associated with a GO term—a gene, a protein, a non-protein-coding RNA, a macromolecular complex, or another gene product. Seven fields of the GAF file specify the annotation object. Each annotation in the GO is associated with a database (field 1) and a database accession number (field 2) that together provide a unique identifier for the gene, the gene product, or the complex. For example, the protein record P00519 is a database object in the UniProtKB database (Fig. 2). The database object symbol (field 3), the database object name (field 10), and the database object synonyms (field 11) provide additional information about the annotation object. The database object type specifies whether the object being annotated is a gene, or a gene product (e.g. protein or RNA; field 12). The organism from which the annotation object is derived is captured as the NCBI taxon ID (taxon; field 13); the corresponding species name can be found at the NCBI taxonomy website.[5]

GO allows capturing isoform-specific data when appropriate; for example UniProtKB accession numbers P00519-1 and P00519-2 are the isoform identifiers for isoform 1 and 2 of P00519. In this case, the database ID still refers to the main isoform, and an isoform accession is included in the GAF file as "Gene Product Form ID" (field 17).

[5] http://www.ncbi.nlm.nih.gov/taxonomy

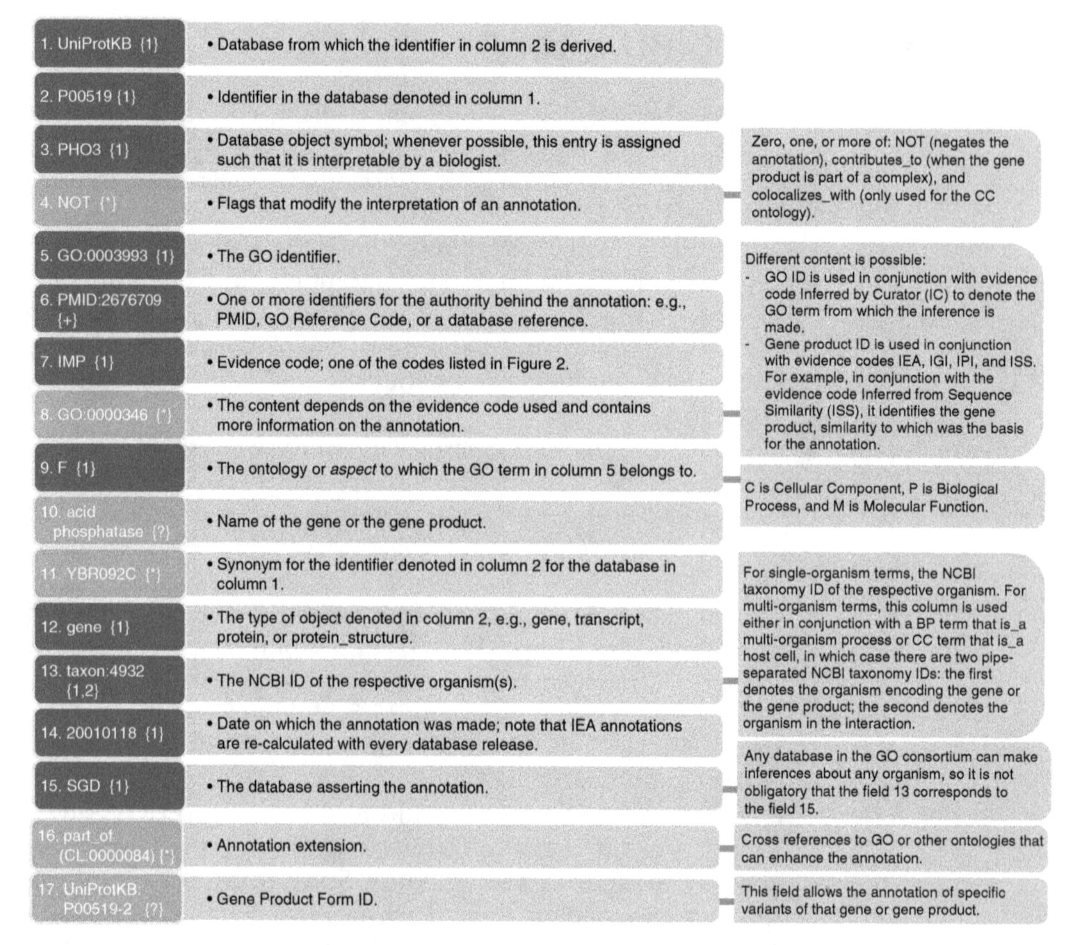

Fig. 2 Gene Association File (GAF) 2.1 file format described with example elements. In the GAF file, each row represents an annotation, consisting of up to 17 tab-delimited fields (or columns). This figure describes these fields in the order in which they are found in the GAF file. *Light blue colour* denotes non-mandatory fields, and these are allowed to be empty in the GAF file. The cardinality—the number of elements in the field—is denoted with the symbol(s) in curly brackets: {?} indicates cardinality of zero or one; {*} indicates that any cardinality is allowed; {+} indicates cardinality of one or more; {1} indicates that cardinality is exactly one; {1,2} indicates that cardinality is either one or two. When cardinality is greater than 1, elements in the field are separated with a pipe character or with a comma; the former indicates "OR" and the latter indicates "AND". The GO term assigned in column 5 is always the most specific GO term possible

5.2 GO Term, Annotation Extension, and Qualifier

Three fields are used to specify the function of the annotation object. Field 5 specifies the GO term, while field 9 denotes the sub-ontology of GO, either Molecular Function, Biological Process, or Cellular Component. While this information is also encoded in the GO hierarchy, explicitly denoting the sub-ontology allows to simplify parsing of the annotations according to the GO aspect. Field 4 denotes the qualifier. One of the three qualifiers can modify the interpretation of an annotation: "contributes_to", "colocalizes_with" and "NOT". This field is not mandatory, but if present it can profoundly change the meaning of an annotation [6]. Thus, while the *producers* of annotations may omit qualifiers, applications that *consume* GO annotations must take them into account. The importance of qualifiers is discussed in more detail in Chap. 14 [14].

An additional field, field 16, is a recent addition to combine more than one term or concept (protein, cell type, etc.) in the same annotation. For example,[6] if a gene product Slp1 is localised to the plasma membrane of T-cells, the GAF file field 16 would contain the information "part_of(CL:0000084 T cell)". Here, CL:0000084 is the identifier for T-cell in the OBO Cell Type (CL) Ontology. This is covered in details in Chap. 17 [28] on annotation extensions.

5.3 Evidence Code and Reference Field

Three fields in the GAF file describe the evidence used to assert the annotation: the Reference (field 6), the Evidence Code (field 7), and the With/From (field 8). The Evidence Code informs the type of experiment or analysis that supports the annotation. There are 21 evidence codes, which can be grouped in three broad categories: experimental annotations, curated non-experimental annotations, and automatically assigned (also known as electronic) annotations (Fig. 3). The Reference field specifies more details on the source of the annotation. For example, when the evidence code denotes an experimentally supported annotation, the Reference will contain the PubMed accession ID (or a DOI if no PubMed ID is available) of the journal article which underpins the annotation, or a GO_REF identifier that refers to a short description of the assignment method, accessible on the GO website.[7] When the evidence code denotes an automatically assigned annotation, i.e. IEA, the reference will contain GO_REF identifiers that specify more details on the automatic assignment, e.g. annotation via the InterPro resource [29].

5.3.1 Experimentally Supported Annotations

Annotations based on direct experimental evidence found in the primary literature are denoted with the general evidence code EXP (Inferred from Experiment) or, when appropriate, the more specific evidence codes IDA (Inferred from Direct Assay), IPI (Inferred from Physical Interaction), IMP (Inferred from Mutant Phenotype), IGI (Inferred from Genetic Interaction), and IEP (Inferred from Expression Pattern) (Fig. 3). These annotations are held in high regard by the community, e.g. [30], and are often used in applications such as checking the enrichment of a gene set in particular functions, finding genes that perform a specific function, or assessing involvement in specific pathways or processes.

[6] http://wiki.geneontology.org/index.php/Annotation_Extension#The_basic_format

[7] http://www.geneontology.org/cgi-bin/references.cgi

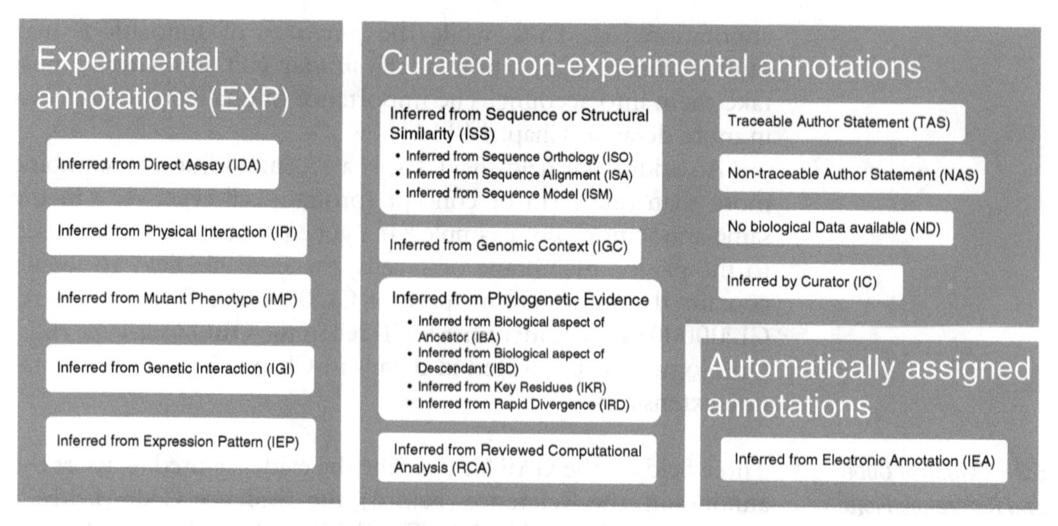

Fig. 3 GO Evidence Codes and their abbreviations. The type of information supporting annotations is recorded with Evidence Codes, which can be grouped into three main categories: experimental evidence codes, curated non-experimental annotations, and automatically assigned annotations. The obsolete evidence code NR (Not Recorded) is not included in the figure. Documentation about the different types of automatically assigned annotations can be found at http://www.geneontology.org/doc/GO.references

Another important use of experimentally supported annotations is in providing trustworthy training sets for various computational methods that infer function [31]. Used this way, the experimentally supported annotations can be amplified to understand more of the growing set of newly sequenced genes.

5.3.2 Curated Non-experimental Annotations

Fourteen of the 21 evidence codes are associated with manually curated non-experimental annotations. Annotations associated with these codes are curated in the sense that every annotation is reviewed by a curator, but they are non-experimental in the sense that there is no direct experimental evidence in the primary literature underpinning them; instead, they are inferred by curators based on different kinds of analyses.

ISS (Inferred from Sequence or Structural Similarity) is a superclass (i.e. a parent) of ISA (Inferred from Sequence Alignment), ISO (Inferred from Sequence Orthology), and ISM (Inferred from Sequence Model) evidence codes. Each of the three subcategories of ISS should be used when only one method was used to make the inference. For example, to improve the accuracy of function propagation by sequence similarity, many methods take into account the evolutionary relationships among genes. Most of these methods rely on orthology (ISO evidence code), because the function of orthologs tends to be more conserved across species than paralogs [32, 33]. In a typical analysis, characterised and uncharacterised genes are clustered based on sequence similarity measures and phylogenetic relationships. The function of unknown genes is then inferred from the function of characterised genes within the same cluster (e.g. [34, 35]).

Another approach to function prediction entails supervised machine learning based on features derived from protein sequence [36–39] (ISM evidence code). Such approach uses a training set of classified sequences to learn features that can be used to infer gene functions. Although few explicit assumptions about the complex relationship between protein sequence and function are required, the results are dependent on the accuracy and completeness of the training data.

IGC (Inferred from Genomic Context) includes, but is not limited to, such things as identity of the genes neighbouring the gene product in question (i.e. synteny), operon structure, and phylogenetic or other whole-genome analysis.

Relatively new are four evidence codes associated with phylogenetic analyses. IBA (Inferred from Biological aspect of Ancestor) and IBD (Inferred from Biological aspect of Descendant) indicate annotations that are propagated along a gene tree. Note that the latter is only applicable to ancestral genes. The loss of an active site, a binding site, or a domain critical for a particular function can be annotated using the IKR (Inferred from Key Residues) evidence code. When this code is assigned by PAINT, GO's Phylogenetic Annotation and INference Tool [40], this means that it is a prediction based on evolutionary neighbours. Finally, negative annotations can be assigned to highly divergent sequences using the code IRD (Inferred from Rapid Divergence).

RCA (inferred from Reviewed Computational Analysis) captures annotations derived from predictions based on computational analyses of large-scale experimental data sets, or based on computational analyses that integrate datasets of several types, including experimental data (e.g. expression data, protein-protein interaction data, genetic interaction data), sequence data (e.g. promoter sequence, sequence-based structural predictions), or mathematical models.

Next, there are two types of annotations derived from author statements. Traceable Author Statement (TAS) refers to papers where the result is cited, but not the original evidence itself, such as review papers. On the other hand a NAS (Non-traceable Author Statement) refers to a statement in a database entry or statements in papers that cannot be traced to another paper.

The final two evidence codes for curated non-experimental annotations are IC (Inferred by Curator) and ND (No biological Data available). If an assignment of a GO term is made using the curator's expert knowledge, concluding from the context of the available data, but without any *direct* evidence available, the IC evidence code is used. For example, if a eukaryotic protein is annotated with the MF term "DNA ligase activity", the curator can assign the BP term "DNA ligation" and CC term "nucleus" with the evidence code IC.

The ND evidence code indicates that the function is currently unknown (i.e. that no characterisation of the gene is currently available). Such an annotation is made to the root of the respective ontology to indicate which functional aspect is unknown. Hence, the ND evidence code allows users for a subtle difference between unannotated genes (for which the literature has not been completely reviewed and thus no GO annotation has been made) and uncharacterised genes (GO annotation with ND code). Note that the ND code is also different from an annotation with the "NOT" qualifier (which indicates the absence of a particular function).

5.3.3 Automatically Assigned Annotations

The evidence code IEA (Inferred from Electronic Annotation) is used for all inferences made without human supervision, regardless of the method used. IEA evidence code is by far the most abundantly used evidence code. The guiding idea behind computational function annotation is the notion that genes with similar sequences or structures are likely to be evolutionarily related, and thus, assuming that they largely kept their ancestral function, they might still have similar functional roles today. For an in-depth discussion of computational methods for GO function annotations, refer to Chap. 5 or *see* refs. [13, 41].

5.3.4 Additional Considerations About Evidence Codes

Biases associated with the different evidence codes are discussed in Chap. 14. Note that there is a more extensive Evidence and Conclusion Ontology (***ECO;*** [42]), formerly known as the "Evidence Code Ontology", presented in Chap. 18 [43] . ECO is only partially implemented in the GO: ECOs are displayed in the AmiGO browser, but they are not in the GAF file. However, all Evidence Codes used by the GO are found also in ECO. There is a general assumption among the GO user community that annotations based on experiments are of higher quality compared to those generated electronically, but this has yet to be empirically demonstrated. Generally, annotations derived from automatic methods tend to be to high-level terms, so they may have a lower information value, but they often withstand scrutiny. Conversely, experiments are sometimes over-interpreted (*see* Chap. 4 [27]) and can also contain inaccuracies.

5.4 Uniqueness of GO Annotations (or Lack Thereof)

No two annotations can have the same combination of the following fields: gene/protein ID, GO term, evidence code, reference, and isoform. Thus one gene can be annotated to the same term with more than one evidence code.

Most GO analyses are gene based, and therefore it is important in such analyses to make sure that the list of genes is non-redundant. However, annotations are often made to larger protein sets that include multiple proteins from the same gene. This is particularly evident in UniProt, which can contain distinct entries from the TrEMBL (unreviewed) portion of the database that do not necessarily represent biologically distinct proteins. The different entries for the same protein or gene are often annotated with identical GO terms, which can bias statistical analyses because some genes have many more entries than other genes. For instance, the set of human proteins in UniProt comprises over 70,000 entries, but there are only approximately 20,000 recognised human protein-coding genes (20,187 reviewed human proteins in the UniProt release of 2015_12). The GO Consortium has worked with UniProt as well as the Quest for Orthologs Consortium to develop "gene-centric" reference proteome lists (http://www.uniprot.org/proteomes/) that provide a single "canonical" UniProt entry for each protein-coding gene. These lists are available for many species, and we encourage users performing gene-centric GO analyses to use only the annotations for UniProt entries in these lists.

6 How Can I Learn More About Gene Ontology Resources?

Most of the topics introduced in this primer will be treated in more depth and nuance in later chapters. Part II focuses on the creation of GO function annotations—we cover in depth the two main strategies of creating GO function annotations: manual extraction/curation from the literature and computational prediction. Part III describes the main strategies used to evaluate their predictive performance. Part IV covers practical uses of the GO annotations: we discuss how GO terms and GO annotations can be summed and compared, how enrichment in specific GO terms can be analysed, and how the GO annotations can be visualised. For the advanced GO user, Part V discusses how the context of a GO annotation is recorded and goes beyond the Evidence Codes to describe how to capture more information on the source of an annotation. We end with Part VI by going beyond GO: we present alternatives to GO for functional annotation; we show how a structured vocabulary is used in the context of controlled clinical terminologies; and we present how information from different structured vocabularies is integrated in one overarching resource.

Acknowledgements

The authors gratefully acknowledge extensive feedback and ideas from Kimberly Van Auken, Marcus C. Chibucos, Prudence Mutowo, and Paul D. Thomas. PG acknowledges National Institutes of Health/National Human Genome Research Institute grant HG002273. CD acknowledges Swiss National Science Foundation grant 150654 and UK BBSRC grant BB/M015009/1. JH acknowledges National Institutes of Health/National Institute for General Medical Sciences grant U24GM088849. Open Access charges were funded by the University College London Library, the Swiss Institute of Bioinformatics, the Agassiz Foundation, and the Foundation for the University of Lausanne.

References

1. Ashburner M, Ball CA, Blake JA et al (2000) Gene ontology: tool for the unification of biology. The Gene Ontology Consortium. Nat Genet 25:25–29
2. Friedmann HC (2004) From "butyribacterium" to "E. coli": an essay on unity in biochemistry. Perspect Biol Med 47:47–66
3. Hastings J (2016) Primer on ontologies. In: Dessimoz C, Škunca N (eds) The gene ontology handbook. Methods in molecular biology, vol 1446. Humana Press. Chapter 1
4. Munoz-Torres M, Carbon S (2016) Get GO! retrieving GO data using AmiGO, QuickGO, API, files, and tools. In: Dessimoz C, Škunca N (eds) The gene ontology handbook. Methods in molecular biology, vol 1446. Humana Press. Chapter 11
5. Carbon S, Ireland A, Mungall CJ et al (2009) AmiGO: online access to ontology and annotation data. Bioinformatics 25:288–289
6. Rhee SY, Wood V, Dolinski K et al (2008) Use and misuse of the gene ontology annotations. Nat Rev Genet 9:509–515
7. Arnaud MB, Costanzo MC, Shah P et al (2009) Gene Ontology and the annotation of pathogen genomes: the case of Candida albicans. Trends Microbiol 17:295–303
8. Pesquita C (2016) Semantic similarity in the gene ontology. In: Dessimoz C, Škunca N (eds) The gene ontology handbook. Methods in molecular biology, vol 1446. Humana Press. Chapter 12
9. Lovering RC, Camon EB, Blake JA et al (2008) Access to immunology through the Gene Ontology. Immunology 125:154–160
10. Bauer S (2016) Gene-category analysis. In: Dessimoz C, Škunca N (eds) The gene ontology handbook. Methods in molecular biology, vol 1446. Humana Press. Chapter 13
11. Burge S, Kelly E, Lonsdale D, et al. (2012) Manual GO annotation of predictive protein signatures: the InterPro approach to GO curation. Database:bar068
12. Pedruzzi I, Rivoire C, Auchincloss AH et al (2015) HAMAP in 2015: updates to the protein family classification and annotation system. Nucleic Acids Res 43:D1064–D1070
13. Cozzetto D, Jones DT (2016) Computational methods for annotation transfers from sequence. In: Dessimoz C, Škunca N (eds) The gene ontology handbook. Methods in molecular biology, vol 1446. Humana Press. Chapter 5
14. Gaudet P, Dessimoz C (2016) Gene ontology: pitfalls, biases, and remedies. In: Dessimoz C, Škunca N (eds) The gene ontology handbook. Methods in molecular biology, vol 1446. Humana Press. Chapter 14
15. Gene Ontology Consortium (2015) Gene Ontology Consortium: going forward. Nucleic Acids Res 43:D1049–D1056
16. The UniProt Consortium (2014) Activities at the Universal Protein Resource (UniProt). Nucleic Acids Res 42:D191–D198
17. Drabkin HJ, Blake JA, Mouse Genome Informatics Database (2012) Manual gene ontology annotation workflow at the Mouse Genome Informatics Database. Database:bas045
18. Hong EL, Balakrishnan R, Dong Q et al (2008) Gene ontology annotations at SGD: new data sources and annotation methods. Nucleic Acids Res 36:D577–D581
19. Davis P, WormBase Consortium (2009) WormBase – nematode biology and genomes. http://precedings.nature.com/documents/3127/version/1/files/npre20093127-1.pdf
20. dos Santos G, Schroeder AJ, Goodman JL et al (2015) FlyBase: introduction of the Drosophila melanogaster Release 6 reference genome assembly and large-scale migration of genome annotations. Nucleic Acids Res 43:D690–D697
21. Gaudet P, Fey P, Basu S et al (2011) dictyBase update 2011: web 2.0 functionality and the initial steps towards a genome portal for the Amoebozoa. Nucleic Acids Res 39:D620–D624
22. Lamesch P, Berardini TZ, Li D et al (2012) The Arabidopsis Information Resource (TAIR): improved gene annotation and new tools. Nucleic Acids Res 40:D1202–D1210
23. Keseler IM, Bonavides-Martínez C, Collado-Vides J et al (2009) EcoCyc: a comprehensive view of Escherichia coli biology. Nucleic Acids Res 37:D464–D470
24. Buchan DWA, Ward SM, Lobley AE et al (2010) Protein annotation and modelling servers at University College London. Nucleic Acids Res 38:W563–W568
25. Burge S, Attwood TK, Bateman A, et al. (2012) Biocurators and biocuration: surveying the 21st century challenges. Database: bar059
26. Lovering RC (2016) How does the scientific community contribute to gene ontology? In: Dessimoz C, Škunca N (eds) The gene ontology handbook. Methods in molecular biology, vol 1446. Humana Press. Chapter 7

27. Poux S, Gaudet P (2016) Best practices in manual annotation with the gene ontology. In: Dessimoz C, Škunca N (eds) The gene ontology handbook. Methods in molecular biology, vol 1446. Humana Press. Chapter 4
28. Huntley RP, Lovering RC (2016) Annotation extensions. In: Dessimoz C, Škunca N (eds) The gene ontology handbook. Methods in molecular biology, vol 1446. Humana Press. Chapter 17
29. Hunter S, Jones P, Mitchell A et al (2012) InterPro in 2011: new developments in the family and domain prediction database. Nucleic Acids Res 40:D306–D312
30. Radivojac P, Clark WT, Oron TR et al (2013) A large-scale evaluation of computational protein function prediction. Nat Methods 10: 221–227
31. The Reference Genome Group of the Gene Ontology Consortium (2009) The Gene Ontology's Reference Genome Project: a unified framework for functional annotation across species. PLoS Comput Biol 5:e1000431
32. Tatusov RL, Koonin EV, Lipman DJ (1997) A genomic perspective on protein families. Science 278:631–637
33. Altenhoff AM, Studer RA, Robinson-Rechavi M et al (2012) Resolving the ortholog conjecture: orthologs tend to be weakly, but significantly, more similar in function than paralogs. PLoS Comput Biol 8:e1002514
34. Mi H, Muruganujan A, Casagrande JT et al (2013) Large-scale gene function analysis with the PANTHER classification system. Nat Protoc 8:1551–1566
35. Altenhoff AM, Škunca N, Glover N et al (2015) The OMA orthology database in 2015: function predictions, better plant support, synteny view and other improvements. Nucleic Acids Res 43:D240–D249
36. Cai C (2003) SVM-Prot: web-based support vector machine software for functional classification of a protein from its primary sequence. Nucleic Acids Res 31:3692–3697
37. Levy ED, Ouzounis CA, Gilks WR et al (2005) Probabilistic annotation of protein sequences based on functional classifications. BMC Bioinformatics 6:302
38. Shen HB, Chou KC (2007) EzyPred: a top-down approach for predicting enzyme functional classes and subclasses. Biochem Biophys Res Commun 364:53–59
39. Lobley AE, Nugent T, Orengo CA et al (2008) FFPred: an integrated feature-based function prediction server for vertebrate proteomes. Nucleic Acids Res 36:W297–W302
40. Gaudet P, Livstone MS, Lewis SE et al (2011) Phylogenetic-based propagation of functional annotations within the Gene Ontology Consortium. Brief Bioinform 12:449–462
41. Škunca N, Altenhoff A, Dessimoz C (2012) Quality of computationally inferred gene ontology annotations. PLoS Comput Biol 8:e1002533
42. Chibucos MC, Mungall CJ, Balakrishnan R et al. (2014) Standardized description of scientific evidence using the Evidence Ontology (ECO). Database:bau075
43. Chibucos MC, Siegele DA, Hu JC, Giglio M (2016) The evidence and conclusion ontology (ECO): supporting GO annotations. In: Dessimoz C, Škunca N (eds) The gene ontology handbook. Methods in molecular biology, vol 1446. Humana Press. Chapter 18

3

Primer on Ontologies

Janna Hastings

Abstract

As molecular biology has increasingly become a data-intensive discipline, ontologies have emerged as an essential computational tool to assist in the organisation, description and analysis of data. Ontologies describe and classify the entities of interest in a scientific domain in a computationally accessible fashion such that algorithms and tools can be developed around them. The technology that underlies ontologies has its roots in logic-based artificial intelligence, allowing for sophisticated automated inference and error detection. This chapter presents a general introduction to modern computational ontologies as they are used in biology.

Key words Ontology, Knowledge representation, Bioinformatics, Artificial intelligence

1 Introduction

Examining aspects of the world to determine the nature of the entities that exist and their causal networks is at the heart of many scientific endeavours, including the modern biological sciences. Advances in technology have made it possible to perform large-scale high-throughput experiments, yielding results for thousands of genes or gene products in single experiments. The data from these experiments are growing in public repositories [1], and in many cases the bottleneck has moved from the generation of these data to the analysis thereof [2]. In addition to the sheer volume of data, as the focus has moved to the investigation of systems as a whole and their perturbations [3], it has become increasingly necessary to integrate data from a variety of disparate technologies, experiments, labs and even across disciplines. Natural language data description is not sufficient to ensure smooth data integration, as natural language allows for multiple words to mean the same thing, and single words to mean multiple things. There are many cases where the meaning of a natural language description is not fully unambiguous. Ontologies have emerged as a key technology going beyond natural language in addressing these challenges.

The most successful biological ontology (bio-ontology) is the Gene Ontology (GO) [4], which is the subject of this volume.

Ontologies are computational structures that describe the entities and relationships of a domain of interest in a structured computable format, which allows for their use in multiple applications [5, 6]. At the heart of any ontology is a set of entities, also called classes, which are arranged into a hierarchy from the general to the specific. Additional information may be captured such as domain-relevant relationships between entities or even complex logical axioms. These entities that are contained in ontologies are then available for use as hubs around which data can be organised, indexed, aggregated and interpreted, across multiple different services, databases and applications [7].

2 Elements of Ontologies

Ontologies consist of several distinct elements, including classes, metadata, relationships, formats and axioms.

2.1 Classes

The class is the basic unit within an ontology, representing a type of thing in a domain of interest, for example *carboxylic acid*, *heart*, *melanoma* and *apoptosis*. Typically, classes are associated with a unique identifier within the ontology's namespace, for example (respectively) CHEBI:33575, FMA:7088, DOID:1909 and GO:0006915. Such identifiers are semantics free (they do not contain a reference to the class name or definition) in order to promote stability even as scientific knowledge and the accompanying ontology representation evolve. Ontology providers commit to maintaining identifiers for the long term, so that if they are used in annotations or other application contexts the user can rely on their resolution. In some cases as the ontology evolves, multiple entries may become merged into one, but in these cases alternate identifiers are still maintained as secondary identifiers. When a class is deemed to no longer be needed within the ontology it may be marked as obsolete, which then indicates that the ID should not be used in further annotations, although it is preserved for historical reasons. Obsolete classes may contain metadata pointing to one or more alternative classes that should be used instead.

2.2 Metadata

Classes are usually associated with annotated textual information—metadata. The metadata associated with classes may include any associated secondary (alternate) identifiers and flags to indicate whether the class has been marked as obsolete. It may also include one or more synonyms; for example the synonyms of *apoptotic process* (a class in the GO) include *cell suicide*, *programmed cell death* and *apoptosis*. It further may include cross references to that class in alternative databases and web resources. For example, many Chemical Entities of Biological Interest (ChEBI) [8] entries contain cross references to the KEGG resource [9], which represents those chemicals in the context of the biological pathways they participate in. Textual comments and examples of intended usage may be annotated. It is very important that each class include a clear definition, which provides enough information to pinpoint the meaning of the class and suggest its appropriate use—sufficiently distinguishing different classes in an ontology so that a user can determine which is the best to use for annotation. The definition of apoptosis offered by the Gene Ontology is as follows:

> A programmed cell death process which begins when a cell receives an internal (e.g. DNA damage) or external signal (e.g. an extracellular death ligand), and proceeds through a series of biochemical events (signaling pathway phase) which trigger an execution phase. The execution phase is the last step of an apoptotic process, and is typically characterized by rounding-up of the cell, retraction of pseudopodes, reduction of cellular volume (pyknosis), chromatin condensation, nuclear fragmentation (karyorrhexis), plasma membrane blebbing and fragmentation of the cell into apoptotic bodies. When the execution phase is completed, the cell has died.

2.3 Relations

Classes are arranged in a hierarchy from the general (high in the hierarchy) to the specific (low in the hierarchy). For example, in ChEBI *carboxylic acid* is classified as a *carbon oxoacid*, which in turn is classified as an *oxoacid*, which in turn is classified as a *hydroxide*, and so on up to the root *chemical entity*, which is the most general term in the structure-based classification branch of the ontology.

Despite the hierarchical organisation, most ontologies are not simple trees. Rather, they are structured as *directed acyclic graphs*. This is because it is possible for classes to have multiple parents in the classification hierarchy, and furthermore ontologies include additional types of relationships between entities other than hierarchical classification (which itself is represented by **is_a** relations). All relations are directed and care must be taken by the ontology editors to ensure that the overall structure of the ontology does not contain cycles, as illustrated in Fig. 1.

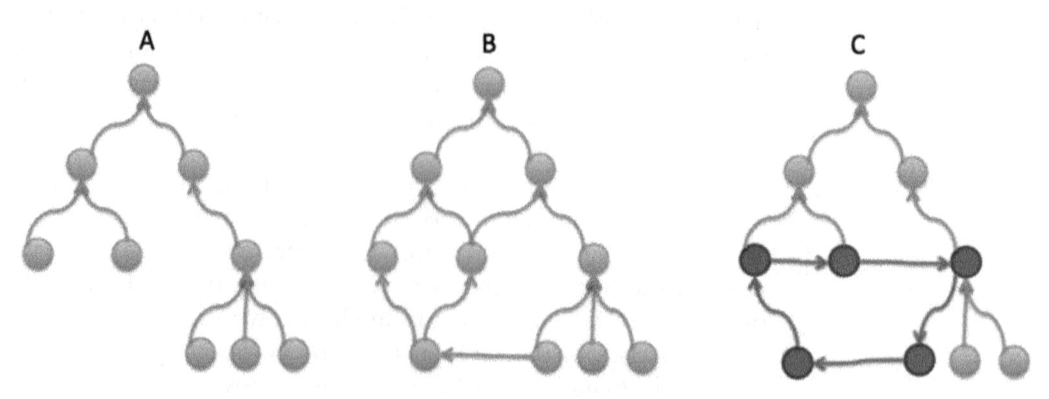

Fig. 1 **(a)** A simple hierarchical tree, **(b)** a directed, acyclic graph, **(c)** a graph that contains a cycle, indicated in *red*

Table 1
A selection of relationship types commonly used in bio-ontologies

Relationship type	Informal meaning	Examples
part_of	The standard relation of parthood.	A brain is part_of a body.
derives_from	Derivation holds between distinct entities when one succeeds the other across a temporal divide in such a way that a biologically significant portion of the matter of the earlier entity is inherited by the latter.	A zygote derives_from a sperm and an ovum.
has_participant	A relation that links processes to the entities that participate in them.	An apoptotic process has_participant a cell.
has_function	A relation that links material entities to their functions, e.g. the biological functions of macromolecules.	An enzyme has_function to catalyse a specific reaction type.

A common relationship type used in multiple ontologies is **part_of** or **has_part**, representing composition or constitution. For example, in the Foundational Model of Anatomy (FMA) [10], *heart* **has_part** *aortic valve*. The Relationship Ontology (RO) defines several relationship types that are commonly used across multiple bio-ontologies [11], a selection of which is shown in Table 1.

In addition, specific ontologies may also include additional relationships that are particular to their domain. For example, GO includes biological process-specific relations such as **regulates**, while ChEBI includes chemistry-specific relationships such as **is_tautomer_of** and **is_enantiomer_of**.

The specification for a relationship type in an ontology includes a unique identifier, name and classification hierarchy, as for classes, as well as a specification whether the relationship is reflexive (i.e. A **rel** B if and only if B **rel** A) and/or transitive (if A **rel** B and B **rel** C then A **rel** C), and the name of the inverse relationship type if it exists. The same metadata as is associated with the classes in the ontology may also be associated with relationship types: alternative identifiers, synonyms, a definition and comments, and a flag to indicate if the relationship is obsolete.

2.4 Formats

Typically, ontologies are stored in files conforming to a specific file format, although there are exceptions that are stored in custom-built infrastructures. Ontologies may be represented in different underlying ontology languages, and historically there has been an evolution of the capability of ontology languages towards greater logical expressivity and complexity, which is mirrored by the advances in computational capacity (hardware) and tools. Biological ontologies such as the GO have historically been represented in the human-readable Open Biomedical Ontologies (OBO) language,[1] which was designed specifically for the structure and metadata content associated with bio-ontologies, but in recent years there has been a move towards the Semantic Web standard Web Ontology Language (OWL)[2] largely due to the latter's adoption within a wider community and expansive tool support. Within OWL, specific standardised annotations are used to encode the metadata content of bio-ontologies as OWL annotations. However, the distinction has become cosmetic to some extent, as tools have been created which are able to interconvert between these languages [12], provided that certain constraints are adhered to.

2.5 Axioms

Within logic-based languages such as OWL, statements in ontologies have a definite logical meaning within a set-based logical theory. Classes have instances as members, and logical axioms define constraints on class definitions that apply to all class members. For example, the statement *carboxylic acid* **is_a** *carbon oxoacid* has the logical meaning that all instances of carboxylic acid are also instances of carbon oxoacid:

$$\forall x : CarboxylicAcid(x) \rightarrow CarbonOxoacid(x)$$

The logical languages underlying ontology technology are collectively called Description Logics [13]—in the plural because there are different variants with different levels of complexity. Some of the different ingredients of logical axioms that are available in the OWL language—quantification, cardinality, logical connectives and negation, disjointness and class equivalence—are explained in Table 2.

Like the carboxylic acid example above, each of these axiom types can be expressed as a logical statement. With these axioms, logic-based ontology reasoners are able to check for errors in an ontology. For example, if a class relation is quantified with 'only' such as the hydrocarbon example given in the table, which in logical language means

$$\forall x \forall y : Hydrocarbon(x) \wedge hasPart(x,y) \leftrightarrow Hydrogen(y) \vee Carbon(y)$$

and then if a subclass of hydrocarbon in the ontology has a **has_part** relation with a target other than a hydrogen or a carbon (e.g. an oxygen):

$$Hydrocarbon(a) \wedge hasPart(a,b) \wedge Oxygen(b)$$

that class will be detected as inconsistent and flagged as such by the reasoner.

---

[1] http://www.cs.man.ac.uk/~horrocks/obo/
[2] http://www.w3.org/TR/owl2-overview/

Table 2
Logical constructs available in the OWL language

Language component	Informal meaning	Examples
Quantification: universal (only) or existential (some)	When specifying relationships between classes, it is necessary to specify a constraint on how the relationship should be interpreted: universal quantification means that for all relationships of that type the target has to belong to the specified class, while existential quantification means that at least one member of the target class must participate in a relationship of that type	*molecule* **has_part** some *atom* *hydrocarbon* **has_part** only (*hydrogen* or *carbon*)
Cardinality: exact, minimum or maximum	It is possible to specify the number of relationships with a given type and target that a class must participate in, or a minimum or maximum number thereof.	*human* **has_part** exactly 2 *leg*
Logical connectives: intersection (and) or union (or)	It is possible to build complex expressions by joining together parts using the standard logical connectives and, or.	*vitamin B* equivalentTo (*thiamin* or *riboflavin* or *niacin* or *pantothenic acid* or *pyridoxine* or *folic acid* or *vitamin B12*)
Negation (not)	In addition to building complex expressions using the logical connectives, it is possible to compose negations.	tailless equivalentTo not (**has_part** some *tail*)
Disjointness of classes	It is possible to specify that classes should not share any members.	*organic* disjointFrom *inorganic*
Equivalence of classes	It is possible to specify that two classes—or class expressions—are logically equivalent, and that they must by definition thus share all their members.	*melanoma* equivalentTo (*skin cancer* and **develops_from** some *melanocyte*)

The end result—an ontology which combines terminological knowledge with complex domain knowledge captured in logical form—is thus amenable to various sophisticated tools which are able to use the captured knowledge to check for errors, derive inferences and support analyses.

3 Tools

Developing a complex computational knowledge base such as a bio-ontology (for example, the Gene Ontology includes 43,980 classes) requires tool support at multiple levels to assist the human knowledge engineers (curators) with their monumental task. For editing ontologies, a commonly used freely available platform is Protégé [14]. Protégé allows the editing of all aspects of an ontology including classes and relationships, logical axioms (in the OWL language) and metadata. Protégé furthermore includes built-in support for the execution of automated reasoners to check for logical errors and for ontology visualisation using various different algorithms. Examples of reasoners that can be used within Protégé are HermiT [15] and Fact++ [16]. For the rapid editing and construction of ontologies, various utilities are available, such as the creation of a large number of classes in a single 'wizard' step. The software is open source and has a pluggable architecture, which allows for custom modular extensions. Protégé is able to open both OBO and OWL files, but it is designed primarily for the OWL language. An alternative editor specific to the OBO language is OBO-Edit [17]. Relative to Protégé, OBO-Edit offers more sophisticated metadata searching and a more intuitive user interface.

To browse, search and navigate within a wide variety of bio-ontologies without installing any software or downloading any files, the BioPortal web platform provides an indispensable resource [18] that is especially important when using terminology from multiple ontologies. Additional browsing interfaces for multiple ontologies include the OLS [19] and OntoBee [20]. Most ontologies are also supported by one or more browsing interfaces specific to that single ontology, and for the Gene Ontology the most commonly used interfaces are AmiGO [21] and QuickGO [22].

Large-scale ontologies such as the GO and ChEBI are often additionally supported by custom-built software tailored to their specific use case, for example embedding the capability to create species-specific 'slims' (subsets of terms of the greatest interest within the ontology for a specific scenario) for the GO, or cheminformatics support for ChEBI. As ontologies are shared across communities of users, an important part of the tool support profile is tools for the community to provide feedback and to submit additional entries to the ontology.

4 Applications

The purposes that are supported by modern bio-ontologies are diverse. The most straightforward application of ontologies is to support the structured annotation of data in a database. Here, ontologies are used to provide unique, stable identifiers—associated to a controlled vocabulary—around which experimental data or manually captured reference information can be gathered [23]. An ontology annotation links a database entry or experimental result to an ontology class identifier, which, being independent of the single database or resource being annotated, is able to be shared across multiple contexts. Without such shared identifiers for biological entities, discrepant ways of referring to entities tend to accumulate—different key words, or synonyms, or variants of identifying labels—which significantly hinders reuse and integration of the relevant data in different contexts.

Secondly, ontologies can serve as a rich source of vocabulary for a domain of interest, providing a dictionary of names, synonyms and interrelationships, thereby facilitating text mining (the automated discovery of knowledge from text) [24], intelligent searching (such as automatic query expansion and synonym searching, an example is described in [25]) and unambiguous identification. When used in multiple independent contexts, such a common vocabulary can become additionally powerful. For example, uniting the representation of biological entities across different model organisms allows common annotations to be aggregated across species [26], which facilitates the translation of results from one organism into another in a fashion essential for the modern accumulation of knowledge in molecular biology. The use of a shared ontology also allows the comparison and translation entities from one discipline to another such as between biology and chemistry [27], enabling interdisciplinary tools that would be impossible computationally without a unified reference vocabulary.

While the above applications would be possible even if ontologies consisted only of controlled vocabularies (standardised sets of vocabulary terms), the real power of ontologies comes with their hierarchical organisation and use of formal inter-entity relationships. Through the hierarchy of the ontology, it is possible to annotate data to the most specific applicable term but then to examine large-scale data in aggregate for patterns at the higher level categories. By centralising the hierarchical organisation in an application-independent ontology, different sources of data can be aggregated to converge as evidence for the same class-level inferences, and complex statistical tools can be built around knowledge bases of ontologies combined with their annotations, which check for over-representation or under-representation of given classes in the context of a given dataset relative to the background of everything that is known [28] (for more information *see* Chap. 13 [29]). The knowledge-based relationships captured in the ontology can be used to assign quantitative measures of similarity between entities that would otherwise lack a quantifiable comparative metric [30] (for more information *see* Chap. 12 [31]). And the relationships between entities can be used to power sophisticated knowledge-based reasoning, such as the inference of which organs, tissues and cells belong to in anatomical contexts [32].

With all these applications in mind, it is no wonder that the number and scope of bio-ontologies have been proliferating over the last decades. The OBO Foundry is a community organisation that offers a web portal in which participating ontologies are listed [33]. The web portal currently lists 137 ontologies, excluding obsolete records. Each of these ontologies has biological relevance and has agreed to abide by several community principles, including providing the ontology under an open license. Examples of these ontologies include ChEBI, the FMA, the Disease Ontology [34] and of course the Gene Ontology which is the topic of this book. In the context of the OBO Foundry, different ontologies are now becoming interrelated through inter-ontology relationships [35], and where there are overlaps in content they are being resolved through community workshops.

5 Limitations

Ontologies are a powerful technology for encoding domain knowledge in computable form in order to drive a multitude of different applications. However, they are not one-stop solutions for all knowledge representation requirements. There are certain limitations to the type of knowledge they can encode and the ways that applications can make use of that encoded knowledge.

Firstly, it is important to bear in mind that ontologies are based on logic. They are good at representing statements that are either true or false (categorical), but they cannot elegantly represent knowledge that is vague, statistical or conditional [36]. Classes that derive their meaning from comparison to a dynamic or conditional group (e.g. *the shortest person in the room*, which may vary widely) are also not possible to represent well within ontologies. It can be difficult to adequately capture knowledge about change over time at the class level, i.e. classes in which the members participate in relationships at one time and not at another, as including a temporal index for each relation would require ternary relations which neither the OBO nor the OWL language support.

Furthermore, although the underlying technology for representation and automated reasoning has advanced a lot in recent years, there are still pragmatic limits to ensure the scalability of the reasoning tools. For this reason, higher order logical statements, non-binary relationships and other complex logical constructs cannot yet be represented and reasoned with in most of the modern ontology languages.

Acknowledgements

The author was supported by the European Molecular Biology Laboratory (EMBL). Open Access charges were funded by the University College London Library, the Swiss Institute of Bioinformatics, the Agassiz Foundation, and the Foundation for the University of Lausanne.

References

1. Marx V (2013) Biology: the big challenges of big data. Nature 498:255–260
2. Holzinger A, Dehmer M, Jurisica I (2014) Knowledge discovery and interactive data mining in bioinformatics – state-of-the-art, future challenges and research directions. BMC Bioinformatics 15(Suppl 6):I1
3. Palsson BO (2015) Systems biology: constraint-based reconstruction and analysis. Cambridge University Press, Cambridge
4. Ashburner M, Ball CA, Blake JA et al (2000) Gene ontology: a tool for the unification of biology. Nat Genet 25:25–29
5. Stevens R, Goble CA, Bechhofer S (2000) Ontology-based knowledge representation for bioinformatics. Brief Bioinform 1(4):398–414
6. Bodenreider O, Stevens R (2006) Bio-ontologies: current trends and future directions. Brief Bioinform 7(3):256–274
7. Hoehdorf R, Schofield PN, Gkoutos GV (2015) The role of ontologies in biological and biomedical research: a functional perspective. *Brief Bioinform* (Advance Access) doi:10.1093/bib/bbv011
8. Hastings J, Owen G, Dekker A, Ennis M, Kale N, Muthukrishnan V, Turner S, Swainston N, Mendes P, Steinbeck C (2015) ChEBI in 2016: improved services and an expanding collection of metabolites. Nucleic Acids Res (advance online access). doi:10.1093/nar/gkv1031
9. Kanehisa M, Goto S, Sato Y, Kawashima M, Furumichi M, Tanabe M (2014) Data, information, knowledge and principle: back to metabolism in KEGG. Nucleic Acids Res 42:D199–D205
10. Golbreich C, Grosjean J, Darmoni SJ (2013) The foundational model of anatomy in OWL 2 and its use. Artif Intell Med 57(2): 119–132
11. Smith B, Ceusters W, Klagges B, Köhler J, Kumar A, Lomax J, Mungall C, Neuhaus F, Rector AL, Rosse C (2005) Relations in biomedical ontologies. Genome Biol 6:R46
12. Tirmizi SH, Aitken S, Moreira DA, Mungall C, Sequeda J, Shah NH, Miranker DP (2011) Mapping between the OBO and OWL ontology languages. J Biomed Semantics 2 (Suppl 1):S3
13. Baader F, Calvanese D, McGuinness D, Nardi D, Patel-Schneider PF (2007) The description logic handbook: theory, implementation and applications, 2nd edn. Cambridge University Press, Cambridge
14. Protégé ontology editor. http://protege.stanford.edu/. Last Accessed Nov 2015
15. Shearer R, Motik B, Horrocks I (2008) HermiT: a highly-efficient OWL reasoner. In Proceedings of the 5th international workshop on owl: experiences and directions, Karlsruhe, Germany, 26–27 October 2008
16. Tsarkov D, Horrocks I (2006) Fact++ description logic reasoner: system description. In Proceedings of the third international joint conference on automated reasoning (IJCAR), pp 292–297
17. Day-Richter J, Harris M, Haendel M, The Gene Ontology OBO-Edit Working Group, Lewis S (2007) OBO-Edit—an ontology editor for biologists. Bioinformatics 23(16): 2198–2200
18. Noy NF, Shah NH, Whetzel PL, Dai B et al (2009) BioPortal: ontologies and integrated data resources at the click of a mouse. Nucleic Acids Res 37(Suppl 2):W170–W173
19. Côté RG, Jones P, Apweiler R, Hermjakob H (2006) The Ontology Lookup Service, a lightweight cross-platform tool for controlled vocabulary queries. BMC Bioinformatics 7:97
20. Xiang Z, Mungall C, Ruttenberg A, He Y (2011) OntoBee: a linked data server and browser for ontology terms. In Proceedings of the 2nd international conference on biomedical ontologies (ICBO), 28–30 July, Buffalo, NY, USA, pp 279–281
21. Carbon S, Ireland A, Mungall C, Shu S, Marshall B, Lewis S, The Amigo Hub and the Web Presence Working Group (2008) AmiGO: online access to ontology and annotation data. Bioinformatics 25(2):288–289
22. Binns D, Dimmer E, Huntley R, Barrell D, O'Donovan C, Apweiler R (2009) QuickGO: a web-based tool for Gene Ontology searching. Bioinformatics 25(22):3045–3046
23. Blake J, Bult C (2006) Beyond the data deluge: data integration and bio-ontologies. J Biomed Inform 39(3):314–320
24. Rebholz-Schuhmann D, Oellrich A, Hoehndorf R (2012) Text-mining solutions for biomedical research: enabling integrative biology. Nat Rev Genet 13:829–839

25. Imam, F, Larson, S, Bandrowski, A, Grethe, J, Gupta A, Martone MA (2012) Maturation of neuroscience information framework: an ontology driven information system for neuroscience. In Proceedings of the formal ontologies in information systems conference, Frontiers in artificial intelligence and applications, vol 239, pp 15–28
26. Huntley RP, Sawford T, Mutowo-Meullenet P, Shypitsyna A, Bonilla C, Martin MJ, O'Donovan C (2015) The GOA Database: gene ontology annotation updates for 2015. Nucleic Acids Res 43(Database issue):D1057–D1063
27. Hill DP, Adams N, Bada M, Batchelor C et al (2013) Dovetailing biology and chemistry: integrating the Gene Ontology with the ChEBI chemical ontology. BMC Genomics 14:513
28. Tipney H, Hunter L (2010) An introduction to effective use of enrichment analysis software. Hum Genomics 4(3):202–206
29. Bauer S (2016) Gene-category analysis. In: Dessimoz C, Škunca N (eds) The gene ontology handbook. Methods in molecular biology, vol 1446. Humana Press. Chapter 13
30. Pesquita C, Faria D, Falcao AO, Lord P, Couto FM (2009) Semantic similarity in biomedical ontologies. PLoS Comput Biol 5(7):e1000443
31. Pesquita C (2016) Semantic similarity in the gene ontology. In: Dessimoz C, Škunca N (eds) The gene ontology handbook. Methods in molecular biology, vol 1446. Humana Press. Chapter 12
32. Osumi-Sutherland D, Reeve S, Mungall CJ, Neuhaus F, Ruttenberg A, Jefferis GS, Armstrong JD (2012) A strategy for building neuroanatomy ontologies. Bioinformatics 28(9):1262–1269
33. Smith B, Ashburner M, Rosse C, Bard J et al (2007) The OBO Foundry: coordinated evolution of ontologies to support biomedical data integration. Nat Biotechnol 25: 1251–1255
34. Kibbe WA, Arze C, Felix V, Mitraka E et al (2015) Disease ontology 2015 update: an expanded and updated database of human diseases for linking biomedical knowledge through disease data. Nucleic Acids Res 43:D1071–D1078
35. Mungall CJ, Bada M, Berardini TZ, Deegan J, Ireland A, Harris MA, Hill DP, Lomax J (2011) Cross-product extensions of the gene ontology. J Biomed Inform 44(1):80–86
36. Schulz S, Stenzhorn H, Boeker M, Smith B (2009) Strengths and limitations of formal ontologies in the biomedical domain. Rev Electron Comun Inf Inov Saude 3(1):31–45

Making Gene Ontology Annotations

4

Text Mining to Support Gene Ontology Curation and Vice Versa

Patrick Ruch

Abstract

In this chapter, we explain how text mining can support the curation of molecular biology databases dealing with protein functions. We also show how curated data can play a disruptive role in the developments of text mining methods. We review a decade of efforts to improve the automatic assignment of Gene Ontology (GO) descriptors, the reference ontology for the characterization of genes and gene products. To illustrate the high potential of this approach, we compare the performances of an automatic text categorizer and show a large improvement of +225 % in both precision and recall on benchmarked data. We argue that automatic text categorization functions can ultimately be embedded into a Question-Answering (QA) system to answer questions related to protein functions. Because GO descriptors can be relatively long and specific, traditional QA systems cannot answer such questions. A new type of QA system, so-called Deep QA which uses machine learning methods trained with curated contents, is thus emerging. Finally, future advances of text mining instruments are directly dependent on the availability of high-quality annotated contents at every curation step. Databases workflows must start recording explicitly all the data they curate and ideally also some of the data they do not curate.

Key words Automatic text categorization, Gene ontology, Data curation, Databases, Data stewardship, Information storage and retrieval

1 Introduction

This chapter attempts to concisely describes the role played by text mining in literature-based curation tasks concerned with the description of protein functions. More specifically, the chapter explores the relationships between the Gene Ontology (GO) and Text Mining.

Subheading 2 introduces the reader to basic concepts of text mining applied to biology. For a more general introduction, the reader may refer to a recent review paper by Zheng et al. [1].

Subheading 3 presents the text mining methods developed to support the assignment of GO descriptors to a gene or a gene product based on the content of some published articles. The section also introduces the methodological framework needed to assess the performances of these systems called automatic text categorizers.

Subheading 4 presents the evolution of results obtained today by GOCat, a GO categorizer, which participated in several BioCreative campaigns.

Finally, Subheading 5 discusses an inverted perspective and shows how GO categorization systems are foundational of a new type of text mining applications, so-called Deep Question-Answering (QA). Given a question, Deep QA engines are able to find answers, which are literally found in no corpus.

Subheading 6 concludes and emphasizes the responsibility of national and international research infrastructures, in establishing virtuous relationships between text mining services and curated databases.

2 State of the Art

This section presents the state of the art in text mining from the point of view of a biocurator, i.e., a person who is maintaining the knowledge stored in gene and protein databases.

2.1 Curation Tasks

In modern molecular biology databases, such as UniProt [2], the content is authored by biologists called biocurators. The work performed by these biologists when they curate a gene or a gene product encompasses a relatively complex set of individual and collaborative tasks [3]. We can separate these tasks into two subsets: sequence annotation—any information added to the sequence such as the existence of isoforms—and functional annotation—any information about the role of the gene or gene product in a given pathway or phenotype. Such a separation is partially artificial because a functional annotation can also establish a relationship between the role of a protein and some sequence positions but it is didactically convenient to adopt such a view.

The primary source of knowledge for genomics and proteomics is the research literature. In the context of biocuration, text mining can be defined as a process aimed at supporting biocurators when they search, read, identify entities, and store the resulting structured knowledge. The developments of benchmarks and metrics to evaluate how automatic text mining systems can help performing these tasks are thus crucial.

BioCreative is a community initiative to periodically evaluates the advances in text mining for biology and biocuration.[1] The forum explored a wide span of tasks with emphasis on named-entity recognition. Named-entity recognition covers a large set of methods that seek to locate and classify textual elements into predefined categories such as the names of persons, organizations, locations, genes, diseases, chemical compounds, etc. Thus, querying PubMed

[1] http://biocreative.sourceforge.net/

Table 1
Comparative curation steps supported by text mining

	[4]	[5]
1	Retrieval	Collection
2	Selection	Triage
3	Reading/Passage retrieval	
4	Entity extraction	Entity indexing
5	Entity normalization	
6		Relationship + evidence annotation
7		Extraction of evidences, e.g., images
8	Feed-back	
9		Check of records

Reference [4] describes the curation task as an iterative process (#8 Feed-back) whereas [6] describes it as a linear process (ending with #9 Check of records). Both descriptions are however consistent. Thus, it is possible to align steps #1, #2, and #4 in Table 1. Step #6 is optional in [4] as the process is regarded as an iterative process. This step is an "intelligent" follow up of the curation task, where already annotated functions/properties should receive less priority in the next Retrieval step. In contrast, steps #3 "Reading/passage retrieval" and #6 "Feed-back" is missed by [6], while the "Extraction of evidences" & "Check of record" is missed by [4] Step #5, i.e., the assignment of unique identifiers to descriptors, in [4] is implicit in step #4 of [6]

with the keywords "biocreative" and "information retrieval" returns 8 PMIDs, whereas 32 PMIDs are returned for the keywords "biocreative" and "named entity" [18th of November 2015].

The general workflow of a curation process supported by text mining instruments commonly comprises 6–9 steps as displayed in Table 1, which is a synthesis inspired by both [6] and [4].

Search is often the first step of a text mining pipeline, although information retrieval has received little attention from bioinformaticians active in Text Mining. Fortunately, information retrieval has been explored by other scientific communities and in particular by information scientists via the TREC (Text Retrieval Conferences) evaluation campaigns, *see* ref. 7 for a general introduction. From 2002 to 2015, molecular biology [8], clinical decision-support [9] and chemistry-related information retrieval [10] challenges have been explored by TREC. Interestingly, large-scale information retrieval studies have consistently shown that named-entity recognition has no or little impact on search effectiveness [11, 12].

2.2 From Basic Search to More Advanced Textual Mining

Beyond information retrieval, more elaborated mining instruments can then be derived. Thus, search engines, which return documents or pointers to documents, are often powered with passage retrieval skills [7], i.e., the ability to highlight a particular sentence, a few phrases, or even a few keywords in a given context.

The enriched representation can help the end-user to decide upon the relevance of the document. If for MEDLINE records, such passage retrieval functionalities are not crucial because an abstract is short enough to be rapidly read by a human, passage retrieval tools become necessary when the search is performed on a collection of full-text articles like for instance in PubMed Central. Within a full-text article, the ability to identify the section where a given set of keywords can be very useful as matching the relevant keywords in a "background" section has a different value than matching them in a "results" section. The latter is likely to be a new statement while the former is likely to be regarded as a well-established knowledge.

2.3 Named-Entity Recognition

Unlike in other scientific or technical fields (finance, high energy physics, etc.), in the biomedical domain, named-entity recognition covers a very large set of entities. Such a richness is well expressed by the content of modern biological databases. Text Mining studies have been published for many of those curation needs, including sequence curation and identification of polymorphisms [13], posttranslational modifications [14], interactions with gene products or metabolites [15], etc. In this context, most studies attempted to develop instruments likely to address a particular set of annotation dimensions, serving the needs of a particular molecular biology database. The focus in such studies is often to design a Graphic User Interfaces and to simplify the curation work by highlighting specific concepts in a dedicated tool [16]. While most of these systems seem exploratory studies, some seem deeply integrated in the curation workflow, as shown by the OntoMate tool designed by the Rat Genome Database [17], the STRING DB for protein–protein interactions or the BioEditor of neXtProt [18].

From an evaluation perspective, the idea is to detect the beginning and the end of an entity and to assign a semantic type to this string. Thus in named-entity recognition, we assume that entity components are textually contiguous. Inherited from early corpus works on information extraction and computational linguistics [19], the goal is to assign a unique semantic category—e.g., Time, Location, and Person—to a string in a text [20].

Semantic categories are virtually infinite but some entities received more attention. Gene, gene products, proteins, species [21, 22], and more recently chemical compounds were significantly more studied than for instance organs, tissues, cell types, cell anatomy, molecular functions, symptoms, or phenotypes [23].

The initial works dealing with the recognition of GO entities were disappointing (Subheading 3.2), which may explain part of the reluctance to address these challenges. We see here one important limitation of named entities: it is easy to detect a one or two words terms into a document, while the recognition of a protein function does require a "deeper" understanding or combination of biological concepts. Indeed a complex GO concept is likely to combine subconcepts belonging to various semantic types, including small molecules, atoms, protein families, as well as biological processes, molecular functions, and cell locations.

2.4 Normalization and Relationship Extraction

In order to compensate for the limitations of named-entity recognition frameworks, two more complementary approaches have been proposed: entity normalization and information (or relationship) extraction.

Normalization can be defined as the process by which a unique semantic identifier is assigned to the recognized entities [24]. The identifiers are available in different resources such as several onto-terminologies or knowledge bases. The assignment of unique identifiers can be relatively difficult in practice due to a linguistic phenomenon called lexical ambiguity. Many strings are lexically ambiguous and therefore can receive more than one identifier depending on the context (e.g., *HIV* could be a disease or a virus). The difficulty is amplified in cascaded lexical ambiguities. Many entities require the extraction of other entities to receive an unambiguous identifier. For instance, the assignment of an accession number to a protein may depend on the recognition of an organism or a cell line somewhere else in the text.

Further, the extraction of relationships requires the recognition of the specific entities, which can be as various as a location, an interaction (binding, coexpression, etc.) [25], an etiology or a temporal marker (cause, trigger, simultaneity, etc.) [26]. For some information extraction tasks such as protein–protein interactions, the normalization and relationship extraction may require first the proper identification of other entities such as the experimental methods (e.g., yeast 2-hybrid) used to generate the prediction. Furthermore, additional information items may be provided such as the scale of the interaction or the confidence in the interaction [27].

To identify GO terms, named-entity recognition and information extraction is insufficient due to two main difficulties: first, the difficulty of defining all (or most) strings describing a given concept; second, the difficulty of defining the string boundaries of a given concept. The parsing of texts to identify GO functions and how they are linked with a given protein demands the development of specific methods.

2.5 Automatic Text Categorization

Automatic text categorization (ATC) can be defined as the assignment of any class or category to any text content. The interested reader can refer to [28], where the author provides a comprehensive introduction to ATC, with a focus on machine learning methods.

In both ATC and in Information Retrieval, documents are regarded as “bag-of-words.” Such a representation is an approximation but it is a powerful and productive simplification. From this bag, where all entities and relationships are treated as flat and independent data, ATC attempts to assign a set of unambiguous descriptors. The set of descriptors can be binary as in triage tasks, where documents can be either classified as relevant for curation or irrelevant, or it can be multiclass. The scale of the problem is one parameter of the model. In some situations, ATC systems do not need to provide a clear split between relevant and irrelevant categories. In particular, when a human is in the loop to control the final descriptor assignment step, ATC systems can provide a ranked list of descriptors, where each rank expresses the confidence score of the ATC system. ATC systems and search engines share here a second common point: compared to named-entity recognition, which is normally not interactive, ATC and Information Retrieval are well suited for human–computer interactions.

3 Methods

With over 40,000 terms—and many more if we account for synonyms—assigning a GO descriptor to a protein based on some published document is formally known as a large multiclass classification problem.

3.1 Automatic Text Categorization

The two basic approaches to solve the GO assignment problem are the following: (1) exploit the lexical similarity between a text and a GO term and its synonyms [29]; (2) use some existing database to train a classifier likely to infer associations beyond string matching. The second approach uses any scalable machine learning techniques to generate a model trained on the Gene Ontology Annotation (GOA) database. Several machine learning strategies have been used but the trade-off between effectiveness, efficiency, and scalability often converges toward an approach called k-Nearest Neighbors (k-NN); *see* also ref. 30.

3.2 Lexical Approaches

Lexical approaches for ATC exploit the similarities between the content of a text and the content of a GO term and its related synonyms [31]. Additional information can be taken into account to augment the categorization power such as the definitions of the GO terms. The ranking functions take into account the frequency of words, their specificity (measured by the "inverse document frequency," the inverse of how many documents contain the word), as well as various positional information (e.g., word order); *see* ref. 32 for a detailed description.

The task is extremely challenging if we consider that some GO terms contain a dozen words, which makes those terms virtually unmatchable in any textual repository. The results of the first BioCreative competition, which was addressing this challenge, were therefore disappointing. The best "high-precision" system achieved an 80 % precision but this system covered less than 20 % of the test sample. In contrast, with a recall close to 80%, the best "high-recall" systems were able to obtain an average precision of 20–30% [33]. At that time, over 10 years ago, such a complex task was consequently regarded are practically out of reach for machines.

3.3 k-Nearest Neighbors

The principle of a k-NN is the following: for an instance X to be classified, the system computes a similarity measure between X and some annotated instances. In a GO categorizer, an instance is typically a PMID annotated with some GO descriptors. Instances on the top of the list are assumed "similar" to X. Experimentally, the value of k must be determined, where k is the number of similar instances (or neighbors), which should be taken into account to assign one or several categories to X.

When considering a full-text article, a particular section in this article, or even a MEDLINE record, it is possible to compute a distance between this section and similar articles in the GOA database because in the curated section of GOA, many GO descriptors are associated with a PMID—those marked up with an EXP evidence code [34]. The computation of the distance between two arbitrary texts can be more or less complex—starting with counting how many words they share—and the determination of the *k* parameters can also be dependent on different empirical features (number of documents in the collection, average size a document, etc.) but the approach is both effective and computationally simple [7]. Moreover, the ability to index a priori all the curated instances makes possible to compute distances efficiently.

The effectiveness of such machine learning algorithms is directly dependent on the volume of curated data. Surprisingly GO categorizers seem not affected by any concept drift, which affects database and data-driven approaches in general. Even old data, i.e., protein annotated with an early version of the GO, seem useful for k-NN approaches [35]. To give a concrete example, consider proteins curated in 2005 with a version of the Gene Ontology and a MEDLINE reports available at that time: it is difficult to understand why a model containing mainly annotations from 2010 to 2014 would outperform a model containing data from 2003 to 2007 using data exactly centered on 2005. While the GO itself has been expanded by at least a factor 4 in the past decade, the consistency of the curation model has remained remarkably stable.

3.4 Properties of Lexical and k-NN Categorizers

In Fig. 1, we show an example output of GOCat [35], which is maintained by my group at the SIB Swiss Institute of Bioinformatics. The same abstract is processed by GOCat using two different types of classification methods: a lexical approach and a k-NN.

In this example, the title of an article ([36]; "Modulation by copper of p53 conformation and sequence-specific DNA binding: role for Cu(II)/Cu(I) redox mechanism") is used as input to contrast the behavior of the two approaches: This reference is used in

#	Score	GO ID	Name
1	1.00	GO:0003677	DNA binding +/-
2	0.77	GO:0043565	sequence-specific DNA binding (synonym sequence specific dna binding) +/-
3	0.31	GO:0070712	RNA cytidine-uridine insertion (synonym rna cu insertion) +/-
4	0.22	GO:0071103	DNA conformation change (synonym dna conformation modification) +/-
5	0.22	GO:0005488	binding +/-
6	0.21	GO:0051982	copper-nicotianamine transmembrane transporter activity (synonym cu-na chelate transporter activity)+/-
7	0.21	GO:0004008	copper-exporting ATPase activity (synonym cu(2+)-exporting atpase activity) +/-
8	0.19	GO:0009455	redox taxis +/-
9	0.19	GO:0016491	oxidoreductase activity (synonym redox activity) +/-
10	0.19	GO:0051776	detection of redox state (synonym redox sensing) +/-
11	0.17	GO:0002039	p53 binding +/-
12	0.16	GO:0005507	copper ion binding (synonym copper binding) +/-
13	0.15	GO:0000393	spliceosomal conformational changes to generate catalytic conformation +/-

#	Score	GO ID	Name
1	1.00	GO:0005507	copper ion binding +/-
2	0.42	GO:0046688	response to copper ion +/-
3	0.22	GO:0008270	zinc ion binding +/-
4	0.21	GO:0003677	DNA binding +/-
5	0.19	GO:0004784	superoxide dismutase activity
6	0.16	GO:0006878	cellular copper ion homeostasis +/-
7	0.13	GO:0035434	copper ion transmembrane transport +/-
8	0.13	GO:0015677	copper ion import +/-
9	0.13	GO:0071280	cellular response to copper ion +/-
10	0.12	GO:0005375	copper ion transmembrane transporter activity +/-
11	0.12	GO:0005886	Plasma membrane
12	0.11	GO:0016531	copper chaperone activity +/-
13	0.10	GO:0055114	oxidation-reduction process
14	0.10	GO:0019430	removal of superoxide radicals
15	0.10	GO:0046914	transition metal ion binding +/-
16	0.10	GO:0006825	copper ion transport +/-
17	0.09	GO:0006801	superoxide metabolic process
18	0.09	GO:0010273	detoxification of copper ion +/-

Fig. 1 Comparative outputs of lexical vs. k-NN versions of GOCat

UniProt to support the assignment of the "copper ion binding" descriptor to *p53*. We see that the lexical system (left panel) is able to assign the descriptor at rank #12, while the k-NN system (right panel) provides the descriptor in position #1.

Finally, we see how both categorizers are also flexible instruments as they basically learn to rank a set of a priori categories. Such systems can easily be used as fully automatic systems—thus taking into account only the top N returned descriptors by setting up an empirical threshold score—or as interactive systems able to display dozens of descriptors including many irrelevant ones, which then can be discarded by the curator.

Today, GO k-NN categorizers do outperform lexical categorizers; however, the behavior of the two systems is complementary. While the latter is potentially able to assign a GO descriptor, which has rarely or never been used to generate an annotation, the former is directly dependent on the quantity of [GO; PMID] pairs available in GOA.

3.5 Inter-annotator Agreement

An important parameter when assessing text mining tools is the development of a ground truth or gold standard. Thus, typically for GO annotation, we assume that the content of curated databases is the absolute reference. This assumption is acceptable from a methodological perspective, as text mining systems need such benchmarks. However, it is worth observing that two curators would not absolutely agree when they assign descriptors, which means that a 100% precision is purely theoretical. Thus, Camon et al. [37] reports that two GO annotators would have an agreement score of about 39–43%. The upper score is achieved when we consider that the assignment of a generic concept instead of a more specific one (children) is counted as an agreement.

4 Today's Performances

Today, GOCat is able to assign a correct descriptor to a given MEDLINE record two times out of three using the BioCreative I benchmark [35], which makes it useful to support functional annotation. Another type of systems, can be used to support complementary tasks of literature exploration (GoPubMed: [38]) or named-entity recognition [39]. While GOCat attempts to assign GO descriptors to any input with the objective to help curating the content of the input, GoPubMed provides a set of facets (Gene Ontology or Medical Subject Headings) to navigate the result of a query submitted to PubMed.

It is worth observing that GO categorizers work best when they assume that the curator is involved in selecting the input papers (performing a triage or selection task as described in Table 1). Such a setting, inherited from the BioCreative competitions, [33, 40] is questionable for at least two reasons: (1) Curators read full-text articles and not only the abstracts—captions and legends seem especially important; (2) The triage task, i.e., the ability to select an article as relevant for curation, could mostly be performed by a machine, provided that fair training data are available. In 2013, the campaign of BioCreative, under the responsibility of the NCBI, revisited the task [41]. The competitors were provided with full-text articles and they were asked not only to return GO descriptors but also to select a subset of sentences. The evaluation was thus more transparent. A small but high-quality annotated sample of full-text papers was provided [42].

The main results from these experiments are the following; *see* ref. 41 for a complete report describing the competition metrics as well as the different systems participating in the challenge. First, the precision of categorization systems improved by about +225% compared to BioCreative 1. Second, the ability to detect all relevant sentences seems less important than being able to select a few high content-bearing sentences. Thus GOCat achieved very competitive results for both recall and precision in GO assignment task, but interestingly the system performed relatively poorly when focusing on the recall of the sentence selection task, *see* Figs. 2 and 3 for

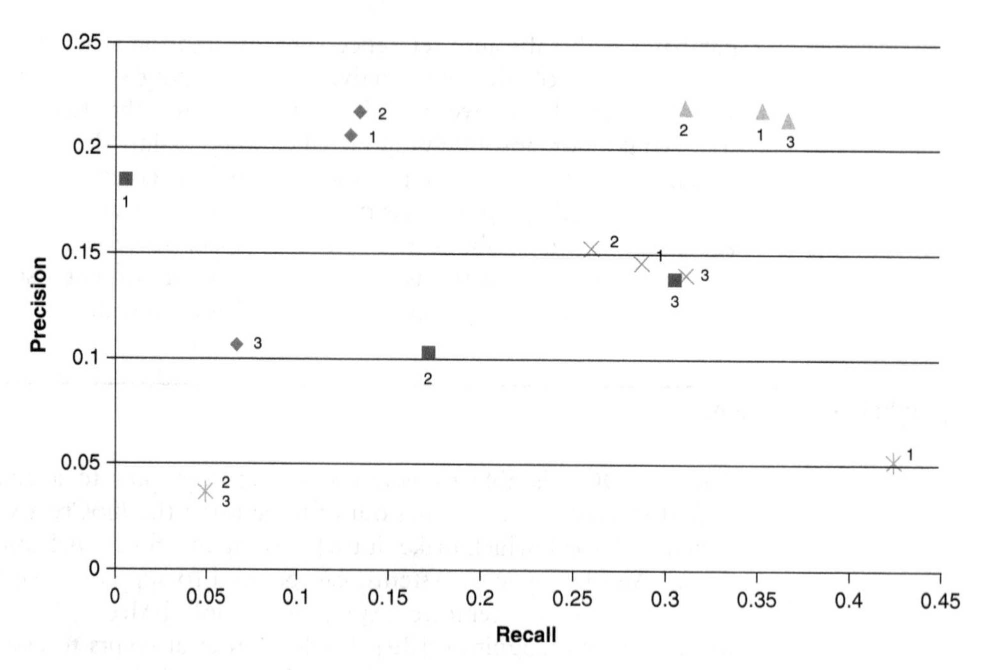

Fig. 2 Relative performance of the sentence triage module of GOCat4FT (GOCat for full-text, *blue diamond*) at the official BioCreative IV competition. Courtesy of Zhiyong Lu, National Institute of Health, National Library of Medicine

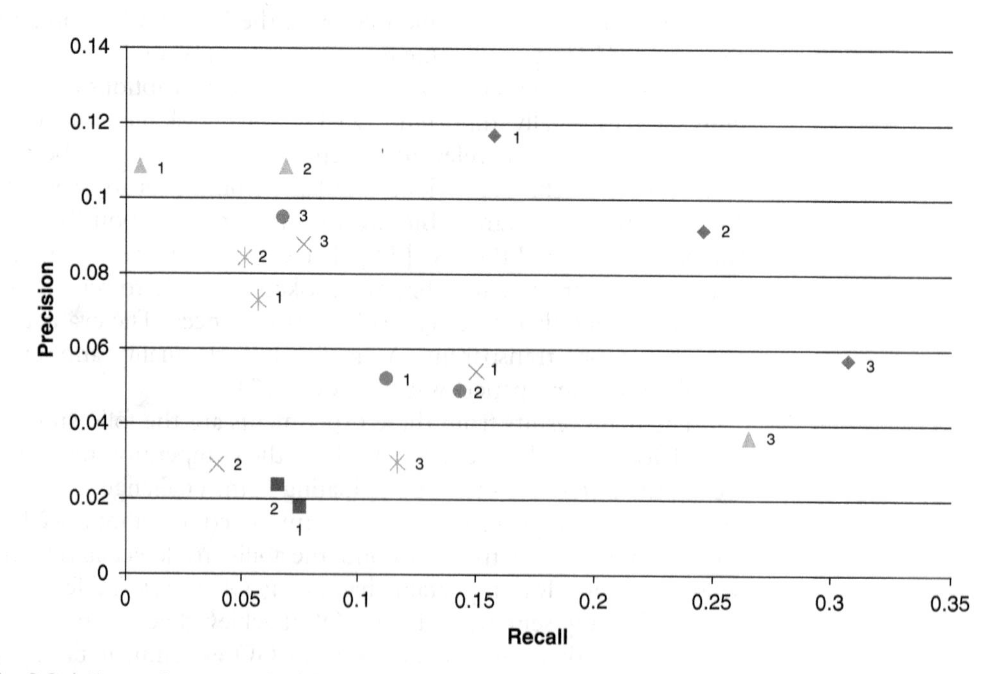

Fig. 3 Relative performance of GOCat4FT (*blue diamond*) when fed with the sentences selected by the three sentence triage systems evaluated in Fig. 2

comparison. We see that two of the sentence ranking systems developed for the BioCreative IV competition (orange dots) outperform other systems in precision but not in recall. References [40, 43] conclude from these experiments that the content in a full-text article is so (highly) redundant that a weak recall is acceptable provided that the few selected sentences have good precision. The few high relevance sentences selected by GOCat4FT (Gene Ontology Categorizer for Full Text) are sufficient to obtain highly competitive results when GO descriptors are assigned by GOCat (orange dots) regarding both recall and precision as the three official runs submitted by SIB Text Mining significantly outperforms other systems. Such a redundancy phenomenon is probably found not only in full-text contents but more generally in the whole literature.

Together with GO and GOA, which was used by most participants in the competition, some online databases seem particularly valuable to help assigning GO descriptors. Thus, Luu et al. [44] uses the cross-product databases [45] with some effectiveness.

5 Discussion

Although a fraction of it is likely to be sufficient to obtain the top-ranked GO descriptors, the results reported in the previous section are obtained by using only 10–20% of the content of an article. This suggests that 80–90% of what is published is unnecessary from an information-theoretic perspective.

5.1 Information Redundancy and Curation-Driven Data Stewardship

New and informative statements are rare in general. They are moreover buried in a mass of relatively redundant and poorly content-bearing claims. It has been shown that the density and precision of information in abstracts is higher [5, 46] than in full-text reports while the level of redundancy across papers and abstracts is probably relatively high as well.

We understand that the separation of valuable scientific statements is labor intensive for curators. This filtering effort is complicated within an article but also between articles at retrieval time. We argue that such task could be performed by machines provided that high-quality training data are available. The training data needed by text mining systems are unfortunately lost during the curation process. Indeed, the separation between useful and useless materials (e.g., PMIDs and sentences) is performed—but not recorded—by the curator during the annotation process but they are unfortunately not stored in databases.

In some cases, the separation is explicit, in other cases, it is implicit but the key point is that a mass of information is definitely lost with no possible recovery. The capture of the output of the selection process—at least for the positive content but ideally also for a fraction of the negative content—is a minimal requirement to improve text mining methods. The expected impact of the implementation of such simple data stewardship recommendation is likely a game changer for text mining far beyond any hypothetical technological advances.

5.2 Assigning Unmatchable GO Descriptors: Toward Deep QA

Some GO concepts describe entities which are so specific that they can hardly be found anywhere. This has several consequences. Traditional QA systems were recently made popular to answer Jeopardy-like questions with entities as various as politicians, town, plants, countries, songs, etc., *see* ref. 47. In the biomedical field, Bauer and Berleant [48] compare four systems, looking at their ergonomics. With a precision in the range of 70–80% [49], these systems perform relatively well. However, none of these systems is able to answer questions about functional proteomics. Indeed, how can a text mining system find an answer if such an answer is not likely to be found on Earth in any corpus of book, article, or patent? The ability to accurately process questions, such as *what molecular functions are associated with tp53* requires to supply answers, such as "RNA polymerase II transcription regulatory region sequence-specific DNA binding transcription factor activity involved in positive regulation of transcription" and only GO categorizers are likely to automatically generate such an answer.

We may think that such complex concepts could be made simpler by splitting the concept into subconcepts, using clinical terminological resources such as SNOMED CT [50, 51] or ICD-10 [52], *see* also Chap. 20 [53]. That might be correct in some rare cases but in general, complex systems tend to be more accurately described using complex concepts. The post-coordination methods explored elsewhere remain effective to perform analytical tasks but they make generative tasks very challenging [52]. Post-coordination is useful to search a database or a digital library because search tasks assume that documents are "bag of words" and they ignore the relationships between these words. However, other tasks such as QA or curation do require to be able to meaningfully combine concepts. In this context, the availability of a pre-computed list of concepts or controlled vocabulary is extremely useful to avoid generating ill-formed entities.

Answering functional omics questions is truly original: it requires the elaboration of a new type of QA engines such as the DeepQA4GO engine [54]. For GO-type of answers, DeepQA4GO is able to answer the expected GO descriptors about two times out of three, compared to one time out of three for traditional systems. We propose to call these new emerging systems: Deep QA engines. Deep QA, like traditional QA engines are able to screen through millions of documents, but since no corpus contain the expected answers, Deep QA is needed to exploit curated biological databases in order to generate useful candidate answers for curators.

6 Conclusion

While the chapter started with introducing the reader to how text mining can support database annotation, the conclusion is that next generation text mining systems will be supported by curated databases. The key challenges have moved from the design of text mining systems to the design of text mining systems able to capitalize on the availability of curated databases. Future advances in text mining to support biocuration and biomedical knowledge discovery are largely in the hands of database providers. Databases workflows must start recording explicitly all the data they curate and ideally also some of the data they do not curate.

In parallel, the accuracy of text mining system to support GO annotation has improved massively from 20 to 65 % (+225 %) from 2005 to 2015. With almost 10,000 queries a month, a tool like GOCat is useful in order to provide a basic functional annotation of protein with unknown and/or uncurated functions [55] as exemplified by the large-scale usage of GOCat by the COMBREX database [56, 57]. However, the integration of text mining support systems into curation workflows remains challenging. As often stated, curation is accurate but does not scale while text mining is not accurate but scales. National and international Research Infrastructures should play a central role to promote optimal data stewardship practices across the databases they support. Similarly, innovative curation models should emerge by combining the quality and richness of curation workflows, more cost-effective crowd-based triage, and the scalability of text mining instruments [58].

References

1. Zeng Z, Shi H, Wu Y, Hong Z (2015) Survey of natural language processing techniques in bioinformatics. Comput Math Methods Med 2015:674296. doi:10.1155/2015/674296, Epub 2015 Oct 7
2. Dimmer EC, Huntley RP, Alam-Faruque Y, Sawford T, O'Donovan C, Martin MJ, Bely B, Browne P, Mun Chan W, Eberhardt R, Gardner M, Laiho K, Legge D, Magrane M, Pichler K, Poggioli D, Sehra H, Auchincloss A, Axelsen K, Blatter MC, Boutet E, Braconi-Quintaje S, Breuza L, Bridge A, Coudert E, Estreicher A, Famiglietti L, Ferro-Rojas S, Feuermann M, Gos A, Gruaz-Gumowski N, Hinz U, Hulo C, James J, Jimenez S, Jungo F, Keller G, Lemercier P, Lieberherr D, Masson P, Moinat M, Pedruzzi I, Poux S, Rivoire C, Roechert B, Schneider M, Stutz A, Sundaram S, Tognolli M, Bougueleret L, Argoud-Puy G, Cusin I, Duek-Roggli P, Xenarios I, Apweiler R (2012) The UniProt-GO Annotation database in 2011. Nucleic Acids Res 40(Database issue):D565–D570. doi:10.1093/nar/gkr1048, Epub 2011 Nov 28
3. Poux S, Magrane M, Arighi CN, Bridge A, O'Donovan C, Laiho K; UniProt Consortium (2014) Expert curation in UniProtKB: a case study on dealing with conflicting and erroneous data. Database (Oxford):bau016. doi: 10.1093/database/bau016
4. Vishnyakova D, Emilie Pasche E, Patrick Ruch P (2012) Using binary classification to priori-

tize and curate articles for the Comparative Toxicogenomics Database. Database 2012

5. Lin J (2009) Is searching full text more effective than searching abstracts? BMC Bioinformatics 10:46. doi:10.1186/1471-2105-10-46
6. Lu Z, Hirschman L. Biocuration workflows and text mining: overview of the BioCreative 2012 Workshop Track II. Database 2012
7. Singhal A (2001) Modern information retrieval: a brief overview. IEEE Data Eng Bull 24:35–43
8. Hersh W, Bhupatiraju RT, Corley S (2004) Enhancing access to the Bibliome: the TREC Genomics Track. Stud Health Technol Inform 107(Pt 2):773–777
9. Simpson MS, Voorhees ES, Hersh W (2014) Overview of the TREC 2014. Clinical Decision Support Track. TREC 2014
10. Lupu M, Huang J, Zhu J, Tait J (2009) TREC-CHEM: large scale chemical information retrieval evaluation at TREC. SIGIR Forum 43(2):63–70
11. Abdou S, Savoy J (2008) Searching in Medline: query expansion and manual indexing evaluation. Inf Process Manag 44(2):781–789
12. Pasche E, Gobeill J, Kreim O, Oezdemir-Zaech F, Vachon T, Lovis C, Ruch P (2014) Development and tuning of an original search engine for patent libraries in medicinal chemistry. BMC Bioinformatics 15(Suppl 1):S15
13. Yip YL, Lachenal N, Pillet V, Veuthey AL (2007) Retrieving mutation-specific information for human proteins in UniProt/Swiss-Prot Knowledgebase. J Bioinform Comput Biol 5(6):1215–1231
14. Veuthey AL, Bridge A, Gobeill J, Ruch P, McEntyre JR, Bougueleret L, Xenarios I (2013) Application of text-mining for updating protein post-translational modification annotation in UniProtKB. BMC Bioinformatics 14:104. doi:10.1186/1471-2105-14-104
15. Xu S, An X, Zhu L, Zhang Y, Zhang H (2015) A CRF-based system for recognizing chemical entity mentions (CEMs) in biomedical literature. J Cheminform 7(Suppl 1 Text mining for chemistry and the CHEMDNER track):S11. doi:10.1186/1758-2946-7-S1-S11. eCollection 2015
16. Dowell KG, McAndrews-Hill MS, Hill DP, Drabkin HJ, Blake JA (2009) Integrating text mining into the MGI biocuration workflow. Database (Oxford):bap019. Epub 2009 Nov 21
17. Liu W, Laulederkind SJ, Hayman GT, Wang SJ, Nigam R, Smith JR, De Pons J, Dwinell MR, Shimoyama M (2015) OntoMate: a text-mining tool aiding curation at the Rat Genome Database. Database (Oxford):bau129
18. SIB Swiss Institute of Bioinformatics Members (2015) The SIB Swiss Institute of Bioinformatics' resources: focus on curated databases. Nucleic Acids Res 44(D1):D27–D37
19. Black WJ, Gilardoni L, Dressel R, Rinaldi F (1997) Integrated text categorisation and information extraction using pattern matching and linguistic processing. RIAO
20. Chinchor N (1997) Overview of MUC-7. Message Understanding Conferences (MUC).
21. Hirschman L, Yeh A, Blaschke C, Valencia A (2005) Overview of BioCreAtIvE: critical assessment of information extraction for biology. BMC Bioinformatics 6(Suppl 1):S1
22. Smith L, Tanabe LK, Ando RJ, Kuo CJ, Chung IF, Hsu CN, Lin YS, Klinger R, Friedrich CM, Ganchev K, Torii M, Liu H, Haddow B, Struble CA, Povinelli RJ, Vlachos A, Baumgartner WA Jr, Hunter L, Carpenter B, Tsai RT, Dai HJ, Liu F, Chen Y, Sun C, Katrenko S, Adriaans P, Blaschke C, Torres R, Neves M, Nakov P, Divoli A, Maña-López M, Mata J, Wilbur WJ (2008) Overview of BioCreative II gene mention recognition. Genome Biol 9(Suppl 2):S2
23. Tran LT, Divita G, Carter ME, Judd J, Samore MH, Gundlapalli AV (2015) Exploiting the UMLS Metathesaurus for extracting and categorizing concepts representing signs and symptoms to anatomically related organ systems. J Biomed Inform. pii:S1532-0464(15)00192-6. doi:10.1016/j.jbi.2015.08.024
24. Morgan AA, Lu Z, Wang X, Cohen AM, Fluck J, Ruch P, Divoli A, Fundel K, Leaman R, Hakenberg J, Sun C, Liu HH, Torres R, Krauthammer M, Lau WW, Liu H, Hsu CN, Schuemie M, Cohen KB, Hirschman L (2008) Overview of BioCreative II gene normalization. Genome Biol 9(Suppl 2):S3. doi:10.1186/gb-2008-9-s2-s3, Epub 2008 Sep 1
25. Bell L, Chowdhary R, Liu JS, Niu X, Zhang J (2011) Integrated bio-entity network: a system for biological knowledge discovery. PLoS One 6(6):e21474
26. Perfetto L, Briganti L, Calderone A, Perpetuini AC, Iannuccelli M, Langone F, Licata L, Marinkovic M, Mattioni A, Pavlidou T, Peluso D, Petrilli LL, Pirrò S, Posca D, Santonico E, Silvestri A, Spada F, Castagnoli L, Cesareni G (2015) SIGNOR: a database of causal relationships between biological entities. Nucleic Acids Res 44:D548–D554

27. Bastian FB, Chibucos MC, Gaudet P, Giglio M, Holliday GL, Huang H, Lewis SE, Niknejad A, Orchard S, Poux S, Skunca N, Robinson-Rechavi M (2015) The Confidence Information Ontology: a step towards a standard for asserting confidence in annotations. Database:bav043 doi:10.1093/database/bav043
28. Sebastiani F (2002) Machine learning in automated text categorization. ACM Comput Surv 34(1):1–47
29. Ruch P (2006) Automatic assignment of biomedical categories: toward a generic approach. Bioinformatics 22(6):658–664, Epub 2005 Nov 15
30. Lena PD, Domeniconi G, Margara L, Moro G (2015) GOTA: GO term annotation of biomedical literature. BMC Bioinformatics 16:346
31. Couto F, Silva M, Coutinho P (2005) FiGO: finding GO terms in unstructured text. BioCreative Workshop Proceedings
32. Ehrler F, Geissbühler A, Jimeno A, Ruch P (2005) Data-poor categorization and passage retrieval for gene ontology annotation in Swiss-Prot. BMC Bioinformatics 6(Suppl 1):S23, Epub 2005 May 24
33. Blaschke C, Leon E, Krallinger M, Valencia A (2005) Evaluation of BioCreAtIvE assessment of task 2. BMC Bioinformatics 6(Suppl 1):S16
34. Gaudet et al. Primer on gene ontology. GO handbook
35. Gobeill J, Pasche E, Vishnyakova D, Ruch P. Managing the data deluge: data-driven GO category assignment improves while complexity of functional annotation increases. Database 2013
36. Hainaut P, Rolley N, Davies M, Milner J (1995) Modulation by copper of p53 conformation and sequence-specific DNA binding: role for Cu(II)/Cu(I) redox mechanism. Oncogene 10(1):27–32
37. Camon EB, Barrell DG, Dimmer EC, Lee V, Magrane M, Maslen J, Binns D, Apweiler R (2005) An evaluation of GO annotation retrieval for BioCreAtIvE and GOA. BMC Bioinformatics 6(Suppl 1):S17, Epub 2005 May 24
38. Doms A, Schroeder M (2005) GoPubMed: exploring PubMed with the Gene Ontology. Nucleic Acids Res 33(Web Server issue):W783–W786
39. Rebholz-Schuhmann D, Arregui M, Gaudan S, Kirsch H, Jimeno A (2008) Text processing through Web services: calling Whatizit. Bioinformatics 24(2):296–298
40. Yeh A, Morgan A, Colosimo M, Hirschman L (2005) BioCreAtIvE task 1A: gene mention finding evaluation. BMC Bioinformatics 6(Suppl 1):S2, Epub 2005 May 24
41. Mao Y, Van Auken K, Li D, Arighi CN, McQuilton P, G Hayman T, Tweedie S, Schaeffer ML, Laulederkind SJF, Wang S-J, Gobeill J, Ruch P, Luu AT, Kim J-J, Chiang J-H, De Chen Y, Yang C-J, Liu H, Zhu D, Li Y, Yu H, Emadzadeh E, Gonzalez G, Chen J-M, Dai H-J, Lu Z (2014). Overview of the gene ontology task at BioCreative IV. Database (Oxford) 2014
42. Van Auken K, Schaeffer ML, McQuilton P, Laulederkind SJ, Li D, Wang SJ, Hayman GT, Tweedie S, Arighi CN, Done J, Müller HM, Sternberg PW, Mao Y, Wei CH, Lu Z (2014) BC4GO: a full-text corpus for the BioCreative IV GO task. Database (Oxford). pii: bau074. doi:10.1093/database/bau074
43. Gobeill J, Pasche E, Dina V, Ruch P. (2014) Closing the loop: from paper to protein annotation using supervised Gene Ontology classification. Database:bau088
44. Luu AT, Kim JJ, Ng SK (2013) Gene ontology concept recognition using cross-products and statistical methods. In: The Fourth BioCreative Challenge Evaluation Workshop, vol. 1, Bethesda, MD, USA, pp 174–181
45. Mungall CJ, Bada M, Berardini TZ et al (2011) Cross-product extensions of the gene ontology. J Biomed Inform 44:80–86
46. Jimeno-Yepes AJ, Plaza L, Mork JG, Aronson AR, Díaz A (2013) MeSH indexing based on automatically generated summaries. BMC Bioinformatics 14:208
47. Ferrucci D (2012) Introduction to « This is Watson ». IBM J Res Dev 56(3.4):1–15
48. Bauer MA, Berleant D (2012) Usability survey of biomedical question answering systems. Hum Genomics 6:17
49. Gobeill J, Patsche E, Teodoro D, Veuthey AL, Lovis C, Ruch P. Question answering for biology and medicine. Information Technology and Applications in Biomedicine, 2009. ITAB 2009
50. Campbell WS, Campbell JR, West WW, McClay JC, Hinrichs SH (2014) Semantic analysis of SNOMED CT for a post-coordinated database of histopathology findings. J Am Med Inform Assoc 21(5): 885–892
51. Dolin RH, Spackman KA, Markwell D (2002) Selective retrieval of pre- and post-coordinated SNOMED concepts. Proc AMIA Symp:210–214
52. Baud RH, Rassinoux AM, Ruch P, Lovis C, Scherrer JR (1999) The power and limits of a

rule-based morpho-semantic parser. Proc AMIA Symp:22–26

53. Denaxas SC (2016) Integrating bio-ontologies and controlled clinical terminologies: from base pairs to bedside phenotypes. In: Dessimoz C, Škunca N (eds) The gene ontology handbook. Methods in molecular biology, vol 1446. Humana Press. Chapter 20
54. Gobeill J, Gaudinat A, Pasche E, Vishnyakova D, Gaudet P, Bairoch A, Ruch P (2015) Deep question answering for protein annotation. Database (Oxford):bav081
55. Mills CL, Beuning PJ, Ondrechen MJ (2015) Biochemical functional predictions for protein structures of unknown or uncertain function. Comput Struct Biotechnol J 13:182–191
56. Anton BP, Chang YC, Brown P, Choi HP, Faller LL, Guleria J, Hu Z, Klitgord N, Levy-Moonshine A, Maksad A, Mazumdar V, McGettrick M, Osmani L, Pokrzywa R, Rachlin J, Swaminathan R, Allen B, Housman G, Monahan C, Rochussen K, Tao K, Bhagwat AS, Brenner SE, Columbus L, de Crécy-Lagard V, Ferguson D, Fomenkov A, Gadda G, Morgan RD, Osterman AL, Rodionov DA, Rodionova IA, Rudd KE, Söll D, Spain J, Xu SY, Bateman A, Blumenthal RM, Bollinger JM, Chang WS, Ferrer M, Friedberg I, Galperin MY, Gobeill J, Haft D, Hunt J, Karp P, Klimke W, Krebs C, Macelis D, Madupu R, Martin MJ, Miller JH, O'Donovan C, Palsson B, Ruch P, Setterdahl A, Sutton G, Tate J, Yakunin A, Tchigvintsev D, Plata G, Hu J, Greiner R, Horn D, Sjölander K, Salzberg SL, Vitkup D, Letovsky S, Segrè D, DeLisi C, Roberts RJ, Steffen M, Kasif S (2013) The COMBREX Project: design, methodology, and initial results. PLoS Biol 11(8):e1001638
57. Škunca N, Roberts RJ, Steffen M (2016) Evaluating computational gene ontology annotations. In: Dessimoz C, Škunca N (eds) The gene ontology handbook. Methods in molecular biology, vol 1446. Humana Press. Chapter 8
58. Burger J, Doughty E, Khare R, Wei CH, Mishra R, Aberdeen J, Tresner-Kirsch D, Wellner B, Kann M, Lu Z, Hirschman L (2014) Hybrid curation of gene-mutation relations combining automated extraction and crowdsourcing. Database (Oxford) 22:2014

5

Best Practices in Manual Annotation with the Gene Ontology

Sylvain Poux and Pascale Gaudet

Abstract

The Gene Ontology (GO) is a framework designed to represent biological knowledge about gene products' biological roles and the cellular location in which they act. Biocuration is a complex process: the body of scientific literature is large and selection of appropriate GO terms can be challenging. Both these issues are compounded by the fact that our understanding of biology is still incomplete; hence it is important to appreciate that GO is inherently an evolving model. In this chapter, we describe how biocurators create GO annotations from experimental findings from research articles. We describe the current best practices for high-quality literature curation and how GO curators succeed in modeling biology using a relatively simple framework. We also highlight a number of difficulties when translating experimental assays into GO annotations.

Key words Gene ontology, Expert curation, Biocuration, Protein annotation

1 Background

Biological databases have become an integral part of the tools researchers use on a daily basis for their work. GO is a controlled vocabulary for the description of biological function, and is used to annotate genes in a large number of genome and protein databases. Its computable structure makes it one of the most widely used resources. Manual annotation with GO involves biocurators, who are trained to reading, extracting, and translating experimental findings from publications into GO terms. Since both the scientific literature and the GO are complex, novice biocurators can make errors or misinterpretations when doing annotation. Here, we present guidelines and recommendations for best practices in manual annotation, to help curators avoid the most common pitfalls. These recommendations should be useful not only to biocurators, but also to users of the GO, since the understanding of the curation process should help understand the meaning of the annotations.

1.1 Knowledge Inference: General Principles

Our understanding of the world is built by observation and experimentation. The overall process of the scientific method involves making hypotheses, deriving predictions from them, and then carrying out experiments to test the validity of these predictions. The results of the experiments are then used to *infer* whether the prediction was true or not [1]. Hypotheses are tested, validated, or rejected, and the combination of all the experiments contributes to uncovering the mechanism underlying the process being studied (Fig. 1).

Examples of experiments include testing an enzymatic activity in vitro using purified reagents, measuring the expression level of a protein upon a given stimulus, or observing the phenotypes of an organism in which a gene has been deleted by molecular genetics techniques. Different inferences can be made from the same experimental setup depending on the hypothesis being tested. Thus, the conclusions that can be derived from individual experiments may vary, depending on a number of factors: they depend on the current state of knowledge, on how well controlled the experiment is, on the experimental conditions, etc. It also happens that the conclusions from a low-resolution experiment are partially or completely refuted when better techniques become available. These factors are inherent to empirical studies and must be taken into account to ensure correct interpretation of experimental results.

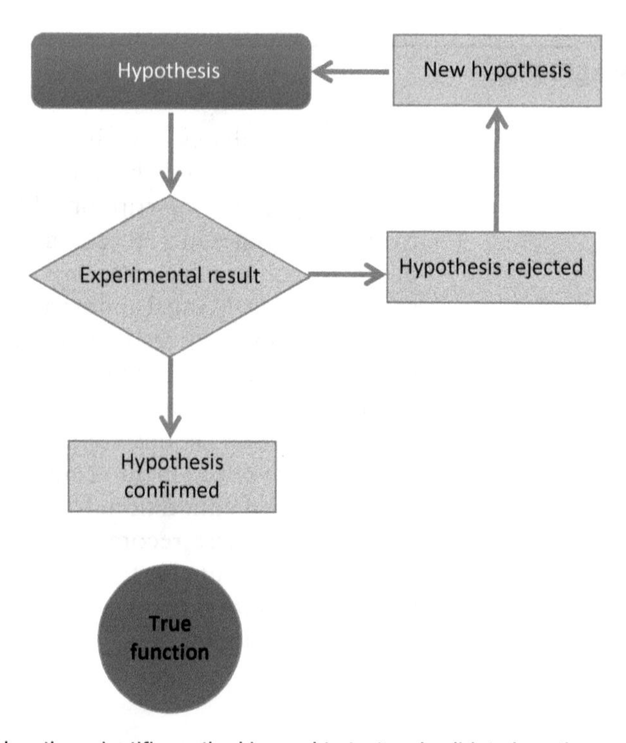

Fig. 1 How the scientific method is used to test and validate hypotheses

1.2 Knowledge Representation Using Ontologies

GO is a framework to describe the roles of gene products across all living organisms [2] (*see* also Chap. 2, [3]). The ontology is divided into three branches, or aspects: Molecular Function (MF) that captures the biochemical or molecular activity of the gene product; Biological Process (BP), corresponding to the wider biological module in which the gene product's MF acts; and Cellular Component (CC), which is the specific cellular localization in which the gene product is active.

The association of a GO term and a gene product is not explicitly defined, but implicitly means that the gene product *has* an activity or a molecular role (MF term), *directly participates* in a process (BP), and the function takes place *in a specific cellular localization* (CC) [2]. Therefore, transient localizations such as endoplasmic reticulum and Golgi apparatus for secreted proteins are not in the scope of GO. Biological process is the most challenging aspect of the GO to capture, in part because it models two categories of processes: *subtypes*: "mitotic DNA replication" (GO:1902969) is a particular type of "nuclear DNA replication" (GO:0033260), and *sub-processes*: mitotic DNA replication is a step of the "cell cycle" (GO:0000278). These two classification axes are distinguished by "is a" and "part of" relations with their parents, respectively. Gene products can be annotated using as many GO terms as necessary to completely describe its function, and the GO terms can be at varying levels in the hierarchy, depending on the evidence available. If a gene product is annotated to any particular term, then the annotations also hold for all the is-a and part-of parent terms. Annotations to more granular terms carry more information; however the annotation cannot be any deeper than what is supported by the evidence.

The complexity of biology is reflected in the GO: with 40,000 different terms [4], learning to use the GO can be compared to learning a new language. As when learning a language, there are terms that are closely related to those we are familiar with, and others that have subtle but important differences in meaning. The GO defines each term in two complementary ways: first by a textual definition intended to be human readable. Secondly, the structure of the ontology as determined by relationships of terms between each other is also a way by which terms are defined these can be utilized for computational reasoning.

1.3 Methods for Assigning GO Annotations

There are two general methods for assigning GO terms to gene products. The first is based on *experimental evidence*, and involves detailed reading of scientific publications to capture knowledge about gene products. Biocurators browse the GO ontologies to associate appropriate GO term(s) whose definition is consistent with the data published for the gene product. *See* Chaps. 3 [5] and 17 [6], for a description of the elements of an annotation. Expert curation based on experiments is considered the gold standard of functional annotation. It is the most reliable and provides strong support for the association of a GO term with a gene product.

The second method involves making *predictions* on the protein's function and subcellular localization, most often with methods relying on sequence similarity. Although not detailed in this chapter, prediction methods are highly dependent on annotations based on experiments. Indeed, all methods to assign annotations based on sequence similarity are more or less directly derived from knowledge that has been acquired experimentally; that is, at least one related protein must have been tested and shown to have a given function for that information to be propagated to other proteins. Hence, the accurate assignment of GO classes to gene products based on experimental results is crucial, since many further annotations depend on their accuracy.

2 Best Practices for High-Quality Manual Curation

2.1 GO Inference Process

Similar to the process by which experimental results get translated into a model of the biological phenomenon being investigated, biocurators take the conclusions from the investigation and convert it into the GO framework. Thus, the same assay may lead to different interpretations depending on the question being tested.

As shown in Table 1, an assay must be interpreted in the wider context of the known roles of the protein, and how directly the assay assesses the protein's role in the process under investigation. Here, several experiments are described in which the readout is DNA fragmentation upon apoptotic stimulation, but that lead to different annotations. DFFB (UniProtKB O76075) is annotated to "apoptotic DNA fragmentation" (GO:0006309) because the protein is also known to be a nuclease. CYCS (UniProtKB P99999) is annotated to caspase activation ("activation of cysteine-type endopeptidase activity involved in apoptotic process" (GO:0006919)) because a direct role has been shown using an in vitro assay. However CYCS is not annotated to "apoptotic DNA fragmentation" (GO:0006309) despite the observation that removing it from cells prevents DNA fragmentation, since the activity of CYCS occurs before DNA fragmentation. Any step that takes place afterwards will inevitably fail to happen, but this does not imply participation in this downstream sequence of molecular events. Finally, the FOXL2 (UniProtKB P58012) transcription factor has a positive effect on the occurrence of apoptosis, by an unknown mechanism, so it is annotated to "positive regulation of apoptotic process" (GO:0043065). This is where the curator's knowledge is critical and provides most added value over, e.g., machine learning and text mining

Table 1
GO inference process, from the hypothesis in the paper to the assay and result, and to the inference of a GO function or role

Protein	Known roles	Hypothesis	Assay → Result	Conclusion → GO	Reference
DDFB (O76075)	DNase	The nuclease activity of DDFB is required for nuclear DNA fragmentation during apoptosis	Apoptotic DNA fragmentation →Increased in the presence of DDFB	DDFB *mediates* nuclear DNA fragmentation during apoptosis →Apoptotic DNA fragmentation (GO:0006309)	[7]
CYCS (P99999)	Cytochrome C; electron transport	CYCS triggers the activation of caspase-3	Apoptotic DNA fragmentation →Decreased upon immunodepletion of CYCS 7 Purified CYCS →Stimulates the auto-proteolytic activity of caspase-3	CYCS *directly activates* caspase-3 →Activation of cysteine-type endopeptidase activity involved in apoptotic process (GO:0006919)	[8]
FOXL2 (P58012)	Transcription factor	Mutations in FOXL2 are known to cause premature ovarian failure, which may be due to increased apoptosis	Apoptotic DNA fragmentation →Increased in the presence of FOXL2	FOXL2 increases the rate of apoptosis →Positive regulation of apoptotic process (GO:0043065)	[9]

2.2 Needles and Haystacks

With more than 500,000 records indexed yearly in PubMed, it is not possible for the GO to comprehensively represent all the available data on every protein. To address this, a careful prioritization of both articles and proteins to annotate is done. The publications from which information is drawn are selected to accurately represent the current state of knowledge. Accessory findings and non-replicated data are not systematically annotated; confirmation or at least consistency with findings from several publications is invaluable to accurately describe the function of a gene product.

Focusing on a topic allows the curator to construct a clear picture of the protein's role and makes it easier to make the best decisions when capturing biological knowledge as annotations. Reading different publications in the field helps to resolve issues and select terms with more confidence. Existing GO annotation in proteins that participate in the same biological process is also helpful to decide on how best to represent the experimental data with the GO. On the other hand, without the broader context of the research domain, some papers may be misleading: first, as more data accumulate, a growing number of contradictory or even incorrect results are found in the scientific literature. Second, the way knowledge evolves occasionally obsoletes previous findings. Curators use their expertise to assess the scientific content of articles and avoid these pitfalls [10].

2.3 How Low Can You Go: Deciding on the Level of Granularity of an Annotation

The level of granularity of an annotation is dictated by the evidence supporting it. A good illustration is provided by ADCK3 protein in human (UniProtKB Q8NI60), an atypical kinase containing a protein kinase domain involved in the biosynthesis of ubiquinone, and an essential lipid-soluble electron transporter. Although it contains a protein kinase domain, it is unclear whether it acts as a protein kinase that phosphorylates other proteins in the CoQ complex or acts as a lipid kinase that phosphorylates a prenyl lipid in the ubiquinone biosynthesis pathway [11]. While it would be tempting to conclude that the protein has "protein kinase activity" (GO:0004672) from the presence of the protein kinase domain, the more general term "kinase activity" (GO:0016301) with no specification of the potential substrate class (lipid or protein) is more appropriate.

2.4 Less Is More: Avoiding Over-Interpretation

2.4.1 Biological Relevance of Experiments

Annotations focus on capturing experiments that are biologically relevant. Thus, substrates, tissue, or cell-type specificity are annotated only when the data indicates the physiological importance of these parameters. One difficulty is that it is not always possible to distinguish between *experimental* context and *biological* context, which can potentially result in GO terms being assigned as if they represented a specific role or under specific conditions, while in fact this only reflects the experimental setup and does not have real biological significance. For example, the activity of E3 ubiquitin protein ligases is commonly tested by an in vitro autoubiquitination assay. While convenient, the assay is not conclusive with respect to the "protein autoubiquitination" (GO:0051865) in vivo. In the absence of additional data, only the term "ubiquitin protein ligase activity" (GO:0061630) should be used. Similarly, the cell type in which a function was tested does not imply that the cell type is relevant for the function; any hint that the protein is studied outside its normal physiological context (such as overexpression) should be carefully taken into consideration.

2.4.2 Downstream Effects

Downstream effects, as well as readouts (discussed above in Subheading 2.1), can lead to incorrect annotations if they are directly assigned to a gene product playing a role many steps further. Here we use downstream as "occurring after," with no implication on the *direct* sequentiality of the events.

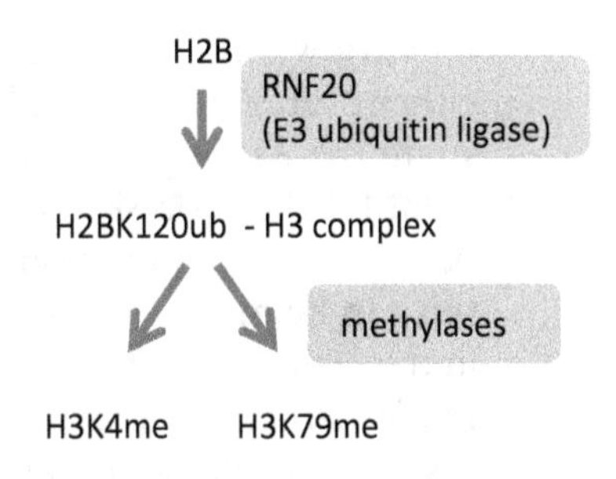

Fig. 2 Monoubiquitination of histone H2B (H2BK120ub) promotes methylation of histone H3 (H3K4me and H3K79me)

Gene products that play housekeeping functions or function upstream of important signaling pathways have many indirect effects and pose a challenge for annotation. This can be illustrated by proteins that mediate chromatin modification. Histone tails are posttranslationally modified by a complex set of interdependent modifications. For instance, histone H2B monoubiquitination at Lys-120 (H2BK120ub) is a prerequisite for the methylation of histone H3 at Lys-4 and Lys-79 (H3K4me and H3K79me, respectively) (Fig. 2). RNF20 (UniProtKB Q5VTR2), an E3 ubiquitin ligase that mediates H2BK120ub, therefore indirectly promotes H3K4me and H3K79me methylation [12]. Thus, the annotation of enzymes that modify histone tails is limited to the primary function of the enzyme ("ubiquitin-protein ligase activity" (GO:0004842) and "histone H2B ubiquitination" (GO:0033523), in this case), while the further histone modifications are only annotated to the proteins mediating these modifications.

A similar approach is taken for cases where the experimental readout is also a GO term. Examples of this include DNA fragmentation assays to measure apoptosis, and MAPK cascade to measure the activation of an upstream pathway. Proteins that are involved in signaling leading to apoptosis do not mediate or *participate* in DNA fragmentation, but their addition or removal causes changes in the amount of DNA fragmentation upon apoptotic stimulation. In other words, the effect of a protein on a specific readout can be very indirect. Whenever possible, annotation of these very specific terms ("apoptotic DNA fragmentation" (GO:0006309), "MAPK cascade" (GO:0000165)) is limited to cases where there is evidence of a molecular function supporting a direct implication in the process. If that information is not available, the annotation is made to a more general term, such as "apoptotic process" (GO:0006915) or "intracellular signal transduction" (GO:0035556), for instance.

2.4.3 Phenotypes

One common method to determine the function or process of a gene is mutagenesis. However, interpreting the results from mutant phenotypes is very difficult, as the effects caused by the absence or disruption of a gene can be very indirect. Any kind of knockout/

knockdown or "add back" experiments (in which proteins are either overexpressed or added to a cellular extract) cannot demonstrate the *participation* of a protein in a process, only its requirement for the process to occur. Inferring a participatory role would be an over-interpretation of the results. A striking illustration of this can be made with housekeeping genes, such as those involved in transcription and translation: knockouts in these proteins (when not lethal) can be pleiotropic and affect essentially all cellular processes. It would be both inaccurate and overwhelming for curators to annotate these gene products to every cellular process impacted. The more prior knowledge we have about a protein's function, in particular its biochemical activity, the more accurate we can be when interpreting a phenotype.

Phenotypes caused by gene mutations are of great interest, not only to try to understand the function of proteins, but also to provide insights into mechanisms leading to disease. The scope of the GO, though, is to capture the *normal* function of proteins. There are phenotype ontologies for human—HPO [13], mouse—MP [14] and other species that allow capturing phenotype in a structure that is more relevant to this type of data.

2.5 Main Functions and Secondary Roles

One limitation of the GO is that main functions and secondary roles are not explicitly encoded, so that this information is difficult to find. For example, enzymes may have different substrates: in some cases, the substrate specificity is driven by the biological context, but in other cases by the experimental conditions. While some activities represent the main function of the enzyme, others are secondary or can be limited to very specific conditions.

A good example is provided by the CYP4F2 enzyme (UniProtKB Q9UIU8), a member of the cytochrome P450 family that oxidizes a variety of structurally unrelated compounds, including steroids, fatty acids, and xenobiotics. In vivo, the enzyme plays a key role in vitamin K catabolism by mediating omega-hydroxylation of vitamin K1 (phylloquinone), and menaquinone-4 (MK-4), a form of vitamin K2 [15, 16]. While hydroxylation of phylloquinone and MK-4 probably constitutes the main activity of this enzyme since this activity has been confirmed by several in vivo assays, CYP4F2 also shows activity towards other related substrates, such as arachidonic acid omega and leukotriene-B [10] omega [17–21]. Clearly vitamin K1 and MK-4 are the main physiological substrates of CYP4F2, but since it is plausible that the enzyme also acts on other molecules, these different activities are also annotated. In the absence of additional evidence, it is currently impossible to highlight which GO term describes the in vivo function of the enzyme. For the reactions known to be implicated in vitamin K catabolism, adding this information as an annotation extension helps clarify the main role of that specific reaction (*see* Chap. 17, [6]).

2.6 Hindsight Is 20/20: Dealing with Evolving Knowledge

Our understanding of biology is dynamic, and evolves as new experiments confirm or contradict previous results. It is therefore essential to read several, preferably recent publications on a subject to make sure that prior working hypotheses, that have subsequently been invalidated, are not annotated. That is, sometimes it is necessary to remove annotations in order to limit the number of false positives. A number of mechanisms exist in GO to capture evolution of knowledge. New GO terms are added to the ontology when knowledge is not covered by existing GO terms. Curators work in collaboration with the GO editors, defining new terms or correcting the definitions of existing terms when required. Conflicting results can be dealt by using the "NOT" qualifier, which states that a gene product is not associated with a GO term. This qualifier is used when a positive association to this term could otherwise be expected from previous literature or automated methods (for more information read www.geneontology.org/GO.annotation.conventions.shtml#not).

A good example of how GO deals with evolving knowledge as new papers are published on a protein is provided by the recent characterization of the NOTUM protein in human and *Drosophila melanogaster*. Notum was first characterized in *D. melanogaster* (UniProtKB Q9VUX3) as an inhibitor of Wnt signaling [22, 23]. Based on its sequence similarity with pectin acetylesterase family members, it was initially thought to hydrolyze glycosaminoglycan (GAG) chains of glypicans by mediating cleavage of their GPI anchor in vitro [24]. Two different articles published recently contradict these previous results, showing that the substrate of human NOTUM (UniProtKB Q6P988) and *D. melanogaster* Notum is not glypicans, and that human NOTUM specifically mediates a palmitoleic acid modification on WNT proteins [25, 26]. This new data confirms the role of NOTUM as an inhibitor of Wnt signaling, but with a mechanism completely different from what the initial studies had suggested. To correctly capture these findings in GO, new terms describing protein depalmitoleylation were added in GO: "palmitoleyl hydrolase activity" (GO:1990699) and "protein depalmitoleylation" (GO:1990697). In addition, NOTUM proteins received negative annotations for "GPI anchor release" (GO:0006507) and "phospholipase activity" (GO:0004620) to indicate that these findings had been disproven.

Although relatively infrequent, this type of situation is critical because it may affect the accuracy of the GO. Ideally, when new findings invalidate previous ones, old annotations are revisited in the light of new knowledge and annotation from previous papers reevaluated to ensure that annotation was not the result of overinterpretation of data.

The most widely used manual protein annotation editor for GO, Protein2GO, has a mechanism to dispute questionable or outdated annotations that sends a request for reevaluation of annotations [27]. Users who notice incorrect or missing annotations are strongly encouraged to notify the GO helpdesk (http://geneontology.org/form/contact-go) so that corrections can be made.

3 Importance of Annotation Consistency: Toward a Quality Control Approach

The goal of the GO project is to provide a uniform schema to describe biological processes mediated by gene products in all cellular organisms [2]. Annotation involves translating conclusions from biological experiments into this schema, such that we are making inferences of inferences. To avoid deriving too much from the biologically relevant conclusions of experiments, consistent annotation within the GO framework is essential.

The GO curators make every effort to ensure that annotations reflect the current state of knowledge. As new findings are made that invalidate or refine existing models there is a need for course correction; otherwise both the ontology and the annotations may drift.

Over 20 groups contribute to manual annotations to the GO project (http://geneontology.org/page/download-annotations). The number of annotations by species, broken down into experimental versus non-experimental, is shown in Fig. 3. Since manual annotations are so critical to the overall quality of the entire corpus of GO data, it is important that each biocurator from every contributing group interprets experiments consistently.

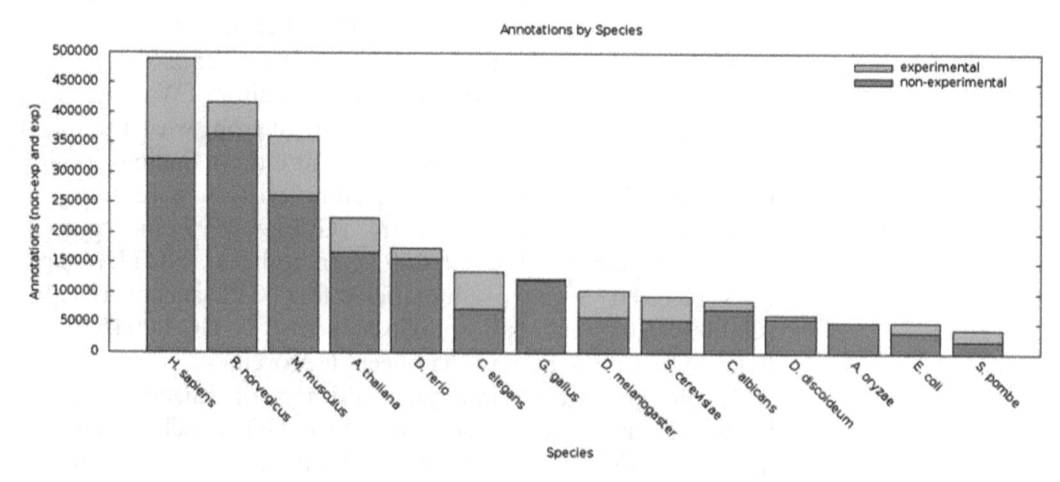

Fig. 3 Number of annotations in 12 species annotated by the GO consortium. Source: http://geneontology.org/page/current-go-statistics

While the GO Consortium does not possess sufficient resources to review all annotations individually on an ongoing basis, several approaches are in place to ensure consistency:

- GO uses automated procedures for validating GO annotations. An automated checker runs through the GO annotation rulebase (http://geneontology.org/page/annotation-quality-control-checks), which validates the syntactic and biological content of the annotation database, and verifies that correct procedures are followed. Examples include taxon checks [28] and checks to ensure that the correct object type is used with different types of evidence.
- The annotation team of the GO consortium also has regular annotation consistency exercises, where participating annotators independently annotate the same paper to ensure that guidelines are applied in a uniform manner, discuss any discrepancy, and update guidelines when these are lacking or need clarification.
- Finally, the Reference Genome Project [29] has proven to be a very useful resource to improve annotation coherence across the GO (Feuermann et al., *in preparation*). The project uses PAINT, a Phylogenetic Annotation and INference Tool, to annotate protein families from the PantherDB resource [30]. PAINT integrates phylogenetic trees, multiple sequence alignments, experimental GO annotations, as well as references pointing to the original data. PAINT curators select the high-confidence data that can be propagated across either the entire tree or specific clades. By displaying different GO annotations for all members of a family, PAINT makes it easy to detect inconsistencies, thus improving the overall quality of the set of GO annotations. It also gives a mean of identifying consistent biases that usually indicate a problem in the ontology or in the annotation guidelines.

4 Summary

Expert curation of GO terms based on experimental data is a complex process that requires a number of skills from biocurators. In this chapter, we describe a number of guidelines to warn curators on common annotation mistakes and provide clues on how to avoid them. These simple rules, summarized in Table 2, can be used as a checklist to ensure that GO annotations are in line with GO consortium guidelines.

Table 2
Summary of annotation guidelines

Carefully select publications. Only annotate papers that provide the most added value.
Read recent publications. Research is not a straightforward process and reading recent publications helps resolving conflicts and detecting experimental discrepancies.
Check annotation consistency. Review the existing annotations for related proteins to see whether the annotations you are adding are consistent.
Look for confirmation for unusual findings with multiple papers, if possible. Avoid entering annotations based on experiments that do not directly implicate the protein with the GO term you annotate.
Annotate the conclusion of the experiment. Keep in mind that this may be different from the results presented. Be especially careful of interpreting the function of proteins based on mutant phenotypes.
Remove obsolete annotations. If you encounter an annotation that is based on an interpretation of an experiment that is no longer valid, use the Challenge mechanism or GO helpdesk to ask to have the annotation removed.

5 Perspective

The guidelines presented here are easy to follow and reinforce curation quality without reducing curation efficiency, which is a serious and valid challenge in the era of big data. In view of the amount of data to be dealt with, it has often been argued that manual curation "just doesn't scale," and an ongoing search for alternative methods is under way in the world of biocuration and bioinformatics. However, examples described in this chapter show that most publications describe complex knowledge that cannot be captured by machine learning or text mining technologies. To continue having an acceptable throughput, manual curation should be able to cope with the increasing corpus of scientific data. From this perspective, PAINT constitutes an excellent example of a propagation tool based on experimental GO annotations, which ensures maximum consistency and efficiency without compromising the quality of the annotations produced. Such system provides one possible answer to the concerns addressed on scalability of expert curation.

References

1. Popper KR (2002) Conjectures and refutations: the growth of scientific knowledge. Routledge, New York
2. Ashburner M, Ball CA, Blake JA, Botstein D, Butler H, Cherry JM, Davis AP, Dolinski K, Dwight SS, Eppig JT et al (2000) Gene ontology: tool for the unification of biology. The Gene Ontology Consortium. Nat Genet 25:25–29
3. Thomas PD (2016) The gene ontology and the meaning of biological function. In: Dessimoz C, Škunca N (eds) The gene ontology handbook. Methods in molecular biology, vol 1446. Humana Press. Chapter 2
4. Gene Consortium (2015) Gene Ontology Consortium: going forward. Nucleic Acids Res 43:D1049–D1056
5. Gaudet P, Škunca N, Hu JC, Dessimoz C (2016) Primer on the gene ontology. In: Dessimoz C, Škunca N (eds) The gene ontology handbook. Methods in molecular biology, vol 1446. Humana Press. Chapter 3
6. Huntley RP, Lovering RC (2016) Annotation extensions. In: Dessimoz C, Škunca N (eds) The gene ontology handbook. Methods in molecular biology, vol 1446. Humana Press. Chapter 17
7. Korn C, Scholz SR, Gimadutdinow O, Lurz R, Pingoud A, Meiss G (2005) Interaction of DNA fragmentation factor (DFF) with DNA reveals an unprecedented mechanism for nuclease inhibition and suggests that DFF can be activated in a DNA-bound state. J Biol Chem 280:6005–6015
8. Liu X, Kim CN, Yang J, Jemmerson R, Wang X (1996) Induction of apoptotic program in cell-free extracts: requirement for dATP and cytochrome c. Cell 86:147–157
9. Lee K, Pisarska MD, Ko JJ, Kang Y, Yoon S, Ryou SM, Cha KY, Bae J (2005) Transcriptional factor FOXL2 interacts with DP103 and induces apoptosis. Biochem Biophys Res Commun 336:876–881
10. Poux S, Magrane M, Arighi CN, Bridge A, O'Donovan C, Laiho K (2014) Expert curation in UniProtKB: a case study on dealing with conflicting and erroneous data. Database (Oxford):bau016
11. Stefely JA, Reidenbach AG, Ulbrich A, Oruganty K, Floyd BJ, Jochem A, Saunders JM, Johnson IE, Minogue CE, Wrobel RL et al (2015) Mitochondrial ADCK3 employs an atypical protein kinase-like fold to enable coenzyme Q biosynthesis. Mol Cell 57:83–94
12. Kim J, Hake SB, Roeder RG (2005) The human homolog of yeast BRE1 functions as a transcriptional coactivator through direct activator interactions. Mol Cell 20:759–770
13. Groza T, Kohler S, Moldenhauer D, Vasilevsky N, Baynam G, Zemojtel T, Schriml LM, Kibbe WA, Schofield PN, Beck T et al (2015) The human phenotype ontology: semantic unification of common and rare disease. Am J Hum Genet 97(1):111–124
14. Smith CL, Eppig JT (2015) Expanding the mammalian phenotype ontology to support automated exchange of high throughput mouse phenotyping data generated by large-scale mouse knockout screens. J Biomed Semantics 6:11
15. Edson KZ, Prasad B, Unadkat JD, Suhara Y, Okano T, Guengerich FP, Rettie AE (2013) Cytochrome P450-dependent catabolism of vitamin K: omega-hydroxylation catalyzed by human CYP4F2 and CYP4F11. Biochemistry 52:8276–8285
16. McDonald MG, Rieder MJ, Nakano M, Hsia CK, Rettie AE (2009) CYP4F2 is a vitamin K1 oxidase: an explanation for altered warfarin dose in carriers of the V433M variant. Mol Pharmacol 75:1337–1346
17. Fava C, Montagnana M, Almgren P, Rosberg L, Lippi G, Hedblad B, Engstrom G, Berglund G, Minuz P, Melander O (2008) The V433M variant of the CYP4F2 is associated with ischemic stroke in male Swedes beyond its effect on blood pressure. Hypertension 52:373–380
18. Jin R, Koop DR, Raucy JL, Lasker JM (1998) Role of human CYP4F2 in hepatic catabolism of the proinflammatory agent leukotriene B4. Arch Biochem Biophys 359:89–98
19. Kikuta Y, Kusunose E, Kondo T, Yamamoto S, Kinoshita H, Kusunose M (1994) Cloning and expression of a novel form of leukotriene B4 omega-hydroxylase from human liver. FEBS Lett 348:70–74
20. Lasker JM, Chen WB, Wolf I, Bloswick BP, Wilson PD, Powell PK (2000) Formation of 20-hydroxyeicosatetraenoic acid, a vasoactive and natriuretic eicosanoid, in human kidney. Role of Cyp4F2 and Cyp4A11. J Biol Chem 275:4118–4126
21. Stec DE, Roman RJ, Flasch A, Rieder MJ (2007) Functional polymorphism in human CYP4F2 decreases 20-HETE production. Physiol Genomics 30:74–81

22. Gerlitz O, Basler K (2002) Wingful, an extracellular feedback inhibitor of Wingless. Genes Dev 16:1055–1059
23. Giraldez AJ, Copley RR, Cohen SM (2002) HSPG modification by the secreted enzyme Notum shapes the Wingless morphogen gradient. Dev Cell 2:667–676
24. Kreuger J, Perez L, Giraldez AJ, Cohen SM (2004) Opposing activities of Dally-like glypican at high and low levels of Wingless morphogen activity. Dev Cell 7:503–512
25. Kakugawa S, Langton PF, Zebisch M, Howell SA, Chang TH, Liu Y, Feizi T, Bineva G, O'Reilly N, Snijders AP et al (2015) Notum deacylates Wnt proteins to suppress signalling activity. Nature 519:187–192
26. Zhang X, Cheong SM, Amado NG, Reis AH, MacDonald BT, Zebisch M, Jones EY, Abreu JG, He X (2015) Notum is required for neural and head induction via Wnt deacylation, oxidation, and inactivation. Dev Cell 32: 719–730
27. Huntley RP, Sawford T, Mutowo-Meullenet P, Shypitsyna A, Bonilla C, Martin MJ, O'Donovan C (2015) The GOA database: gene ontology annotation updates for 2015. Nucleic Acids Res 43:D1057–D1063
28. Deegan (née Clark) JI, Dimmer EC, Mungall CJ (2010) Formalization of taxon-based constraints to detect inconsistencies in annotation and ontology development. BMC Bioinformatics 11:530
29. Gaudet P, Livstone MS, Lewis SE, Thomas PD (2011) Phylogenetic-based propagation of functional annotations within the Gene Ontology consortium. Brief Bioinform 12:449–462
30. Mi H, Muruganujan A, Thomas PD (2013) PANTHER in 2013: modeling the evolution of gene function, and other gene attributes, in the context of phylogenetic trees. Nucleic Acids Res 41:D377–D386

6

How does the Scientific Community Contribute to Gene Ontology?

Ruth C. Lovering

Abstract

Collaborations between the scientific community and members of the Gene Ontology (GO) Consortium have led to an increase in the number and specificity of GO terms, as well as increasing the number of GO annotations. A variety of approaches have been taken to encourage research scientists to contribute to the GO, but the success of these approaches has been variable. This chapter reviews both the successes and failures of engaging the scientific community in GO development and annotation, as well as, providing motivation and advice to encourage individual researchers to contribute to GO.

Key words Clinical and basic research, Gene Ontology, Proteomics, Transcriptomics, Community, Community annotation, Community curation, Genomics, Bioinformatics, Curation, Annotation, Biocuration

1 Introduction

The overarching vision of the Gene Ontology Consortium (GOC) is to describe gene products across species—their temporally and spatially characteristic expression and localization, their contribution to multicomponent complexes, and their biochemical, physiological, or structural functions—and thus enable biologists to easily explore the universe of genomes [1]. In practical terms, this makes providing an accessible, navigable resource of gene products, rigorously described according a structured ontology, the GOC's key objective. The referenced links, between the identifiers for Gene Ontology (GO) terms and the identifiers for specific gene products, are the elemental GO annotations.

With Next Generation Sequencing technologies increasing the rate at which genomic and transcriptomic data are accumulating, the need for highly informative annotation data for the human genome is paramount. Community annotation has the potential to improve the information provided by the GO resource. Consequently, the GOC actively encourages contributions from the scientific community, to ensure that the ontology appropriately reflects the current understanding of biology and to supply gene product annotations [2–4]. There are many online resources that encourage community annotation [5–7]; however, annotations created in the majority of these are not submitted to the GO database. This chapter, therefore, only discusses the progress of community contributions to the GO database.

2 Ontology Development Workshops

The success of GO is dependent on its ability to represent the research communities' interpretation of biological processes and individual gene product functions and cellular locations. This is achieved through the use of descriptive GO terms, with detailed definitions, and appropriate placement of GO terms within the ontology hierarchy. The majority of GO terms are created by GO editors, following a review of the current scientific literature, often, without the need of discussions with experts in the relevant field [8–9].

Major revisions or expansions of a specific GO domain are usually undertaken in consultation with experts working in that biological field. Notable successful ontology development projects include that of the immune system [10], heart development [2], kidney development [11], muscle processes and cellular components [12], cell cycle, and transcription [13]. The expansion of the heart development domain provides a good example of how experts in the field can guide the GO editors to create very descriptive terms. The GO heart development domain describes heart morphogenesis, the differentiation of specific cardiac cell types, and the involvement of signaling pathways in heart development. This was achieved following a 1½ day meeting with four heart development experts, as well as considerable email exchanges both before and after the meeting [2]. The result of this effort was an increase in the number of GO terms describing heart development from 12 to over 280, and the creation of highly expressive terms such as secondary heart field specification (GO:0003139) and canonical Wnt signaling in cardiac neural crest cell differentiation (GO:0061310).

3 Community Contributions to the GO Annotation Database

Lincoln Stein suggested that there are four organizational models to genome annotation: the factory (reliant on a high degree of automation), the museum (requiring expert curators), the cottage industry (scientists working out of their laboratories), and the party (or jamboree—a short intensive annotation workshop) [14]. To this, list needs to be added "the school," where people are encouraged to annotate as part of a bioinformatics training program.

Currently, there are two major approaches taken to associate GO terms with gene products: manual curation of the literature and automated pipelines based on manually created rules (the "factory") [15]. The majority of manual annotation follows the "museum" model, relying on highly trained curators reading the published literature, evaluating the experimental evidence, and applying the appropriate GO terms to the gene record [8, 16]. The majority of these curators are associated with specific model organism databases, such as FlyBase [17], PomBase [18] and ZFIN [19], or proteomic databases, such as UniProt [20]. In general, these curators will be annotating gene products across a whole genome. In contrast, there have been a few annotation projects funded to improve the representation of specific biological domains, such as cardiovascular [3], kidney [21] and neurological [22]. Two of these projects are being undertaken by the UCL functional annotation team and provide an example of an expert curation team embedded within a scientific research group.

3.1 GO Annotation Within a Bioinformatics Course

In the "school" model, bioinformatics courses, which include an introduction to GO, provide an opportunity for attendees to contribute GO annotations. However, providing timely feedback to degree students is very labor intensive. Texas A&M University has circumvented this problem through the use of competitive peer review. A biannual multinational student competition has been established to undertake large-scale manual annotation of gene function using GO. In this competition, known as the Community Assessment of Community Annotation with Ontologies (CACAO),[1] teams of students get points for making annotations, but can also take points from competitors by correcting their annotations. A professional curator then reviews these and annotations that are judged to be correct are submitted to the GO database. This highly successful crowd-source project uses the online GONUTs wiki [23] to submit annotations and has supplied 3700 annotations to the GO database. The CACAO attribution identifies the resultant annotations, associated with over 2500 proteins. This competition has given over 700 students the opportunity not only to learn how to use some of the essential online biological knowledgebases, but to reinforce this knowledge over a 3-month period, connecting their curriculum to research applications. An MSc literature review project, at University College London (UCL), also provides an opportunity to supply GO annotations to the GO database. Four projects, to date, have resulted in annotations for proteins involved in autism [24], heart development, folic acid metabolism, and hereditary hemochromatosis, creating over 1000 annotations. A limitation of student annotations is that they do not draw on the expertise of the scientific community.

[1] http://gowiki.tamu.edu/wiki/index.php/Category:CACAO

For the past 5 years, the UCL functional annotation team has run a 2-day introduction to bioinformatics and GO course. This course has been attended by over 200 scientists, who have been given the opportunity to use the UniProt GO annotation tool, Protein2GO [20], to annotate their own papers or those published in their field of expertise. However, on average only 50 annotations are submitted during the entire course and very few scientists continue to contribute annotations after the end of the course. A similar problem has been identified in many other annotation workshops.

3.2 Annotation Workshops

The first workshop to submit GO annotations to the GO database focused on the annotation of the *Drosophila* genome [25]. Following on from this, the Pathema group ran several annotation-training workshops, in 2007, with the idea that trained scientists would continue to provide annotation updates thereafter [26]. Unfortunately, this approach had limited success. Although 150 scientists attended, in general they provided guidance to the curators, rather than creating annotations themselves.

3.3 GO Annotation by Specific Scientific Communities

One of the most successful community annotation projects is that run by PomBase [18]. During pilot projects, PomBase encouraged 80 scientists from the fission yeast community to submit a variety of annotations, including 226 GO annotations,[2] using their curation tool, CANTO [4]. Following on from this success the PomBase team now receives regular annotations from the *Schizosaccharomyces pombe* community.

Another successful community annotation project has a transcription focus and was initiated by a group at the Norwegian University of Science and Technology. To ensure a consistent annotation approach is undertaken, the Norwegian research group, with members of the GOC, has created a set of transcription factor annotation guidelines [13]. These provide details of the ideal GO terms to associate with a transcription factor, with a list of experimental conditions that would support these annotations. By using these standardized conventions, the literature-curated data (currently including annotations for 400 proteins) is imported directly into the GO database, with only minimal quality checking required. Working with the GOC, the SYSCILIA consortium may prove to be just as effective. This group has already contributed to the development of GO terms to describe ciliary components and processes and started to submit GO annotations [27].

The outstanding contributions of Ralf Stephan, demonstrates what can be achieved through dedication.[3] Stephan singlehandedly annotated 60% of the *Mycobacterium tuberculosis* genome, through the review of over 1000 papers. Furthermore, the resultant 7700

[2] http://www.pombase.org/community/fission-yeast-community-curation-pilot-project

[3] http://www.ark.in-berlin.de/Site/MTB-GOA.html

annotations associated with 2500 proteins were checked by the UniProt-GOA team [15] and needed very few edits, before incorporation into the GO database.

The success of PomBase may reflect the small size of the research community and that an early visionary investment has had a significant impact on the quality of data available at PomBase, achieved through the contributions of individual scientists and curators. In contrast, the Norwegian transcription factor project, formed to address the deficit of transcription factor annotations and in response to a need for comprehensive annotation of these proteins. The creation of a comprehensive and detailed annotation guide is key to the achievements of this project [13]. However, the GO database would also benefit from a few more "cottage industry" contributions, such as those provided for the *Mycobacterium tuberculosis* genome.

4 Why Contribute to GO?

The motivation behind "community annotation" is varied. Some scientists are contributing GO annotations purely to ensure their research area or gene product(s) of interest are well curated. Others may want to ensure data from their own papers is curated and, therefore, promoted in popular knowledgebases; potentially increasing the citation rate of these papers. Others still are motivated by peer competition! Regardless of the motivation, the GOC is always appreciative of input from the scientific community. Despite the success of some community annotation projects, taken as a whole, very few scientists suggest annotations, or papers for annotation. Consequently, the GOC continues to search for new ways to encourage the research community to contribute to curation activities. For example, the inclusion of data from gene wikis [5–7] could help take community annotation forwards. Considerable funding is being invested in NGS, proteomic and transcriptomic technologies and sequencing of population genomes. However, comprehensive gene annotation is likely to be a limiting factor in the identification of genes involved in polygenic diseases and disease-associated disregulated pathways. Many groups are turning to proprietary resources to provide these annotations [28], which also include freely available annotation data. A more sustainable approach, and one that will also support genomic research in developing countries, is to invest in improving the freely available annotation resources. All groups working with high-throughput datasets should consider working with the GOC and including in grant applications a component that would fund the submission of gene annotation data describing their area of interest, by expert curators, rather than requesting funding to enable access to proprietary software. The majority of members of the GOC do provide facilities to enable researchers to contribute to GO, the question is whether the scientific community will acknowledge that their input is required.

5 Resources Supporting Expert Contributions to GO

It is unrealistic to expect a limited number of GO curators and editors to understand all areas of biological and medical research. Consequently, a range of online facilities have been put in place to encourage scientists to review the ontology, to comment on the annotations, and to suggest papers for curation. In addition, several GO annotation tools, enable scientists to contribute annotation data [4, 8, 20]. Furthermore, the Protein2GO curation tool, automatically emails authors when one of their papers has been annotated, giving the authors an opportunity to comment on the curator's interpretation of their data [20].

Scientists interested in helping to improve the GO annotation resource can either contact the group providing annotations to their species or area of interest (see GOC contributors webpage geneontology.org/page/go-consortium-contributors-list) or submit enquires or information through the GOC webform geneontology.org/form/contact-go, which will be forwarded to the relevant database or group. Useful information to provide would be: details of key experimental publications for curation; a review of a particular annotation set (associated with a specific gene product or GO term), pointing out GO annotations that are missing, wrong, or controversial; comments on the ontology structure or definitions of GO terms, with a reference to support the changes required (Fig. 1). This would ensure that any erroneous annotations are removed promptly from the GO database, and that information from seminal papers is included. Scientists who are confident in using online resources may prefer to submit GO annotations, for any species, using the PomBase curation tool, CANTO curation.pombase.org/pombe [4]. Information provided by any of these means will be forwarded to the appropriate curation or editorial team and contributors will be notified when their suggestions have been incorporated. Full details about contributing to GO are available on the GOC website http://geneontology.org/page/contributing-go. Professional GO curators review all submitted annotations to ensure the annotations follow GO annotation rules and a consistent annotation approach is taken.

6 Following GO Developments

Scientists interested in finding out more about current GOC annotation and ontology development projects should sign up to the go-friends mailing list.[4] Alternatively, GO-relevant tweets can be followed via #geneontology, or @news4GO.

[4] http://mailman.stanford.edu/mailman/listinfo/go-friends

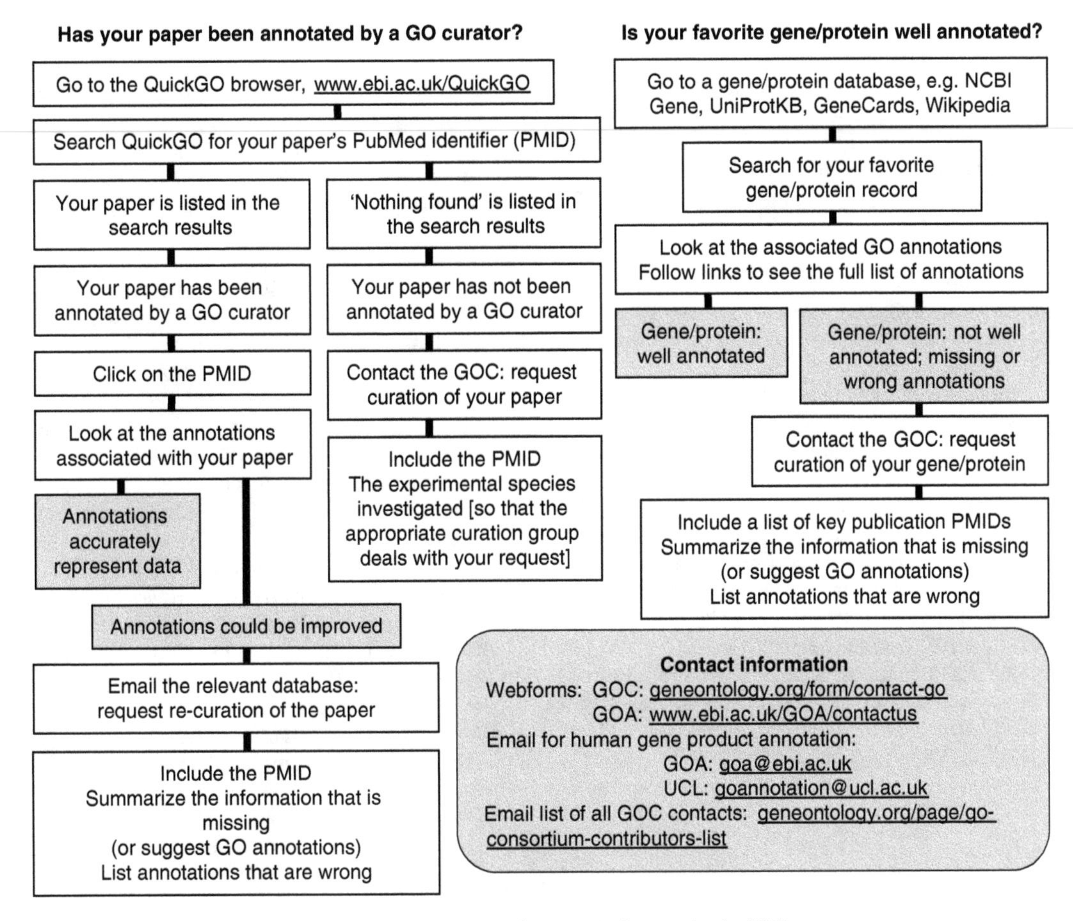

Fig. 1 How research scientists can help to improve the annotation content of GO

Acknowledgments

Supported by the British Heart Foundation (RG/13/5/30112), Parkinson's UK (G-1307), and the National Institute for Health Research University College London Hospitals Biomedical Research Centre. Many thanks to Dr. Rachael Huntley and Professor Suzanna Lewis for their reviews of this manuscript and to Doug Howe and Tanya Berardini for the information they provided. Open Access charges were funded by the University College London Library, the Swiss Institute of Bioinformatics, the Agassiz Foundation, and the Foundation for the University of Lausanne.

References

1. Ashburner M, Ball CA, Blake JA, Botstein D, Butler H et al (2000) Gene ontology: tool for the unification of biology. The Gene Ontology Consortium. Nat Genet 25:25–29
2. Khodiyar VK, Hill DP, Howe D, Berardini TZ, Tweedie S et al (2011) The representation of heart development in the gene ontology. Dev Biol 354:9–17
3. Lovering RC, Dimmer EC, Talmud PJ (2009) Improvements to cardiovascular gene ontology. Atherosclerosis 205:9–14
4. Rutherford KM, Harris MA, Lock A, Oliver SG, Wood V (2014) Canto: an online tool for community literature curation. Bioinformatics 30:1791–1792
5. Singh M, Bhartiya D, Maini J, Sharma M, Singh AR et al (2014) The Zebrafish GenomeWiki: a crowdsourcing approach to connect the long tail for zebrafish gene annotation. Database (Oxford) 2014:bau011
6. Huss JW 3rd, Orozco C, Goodale J, Wu C, Batalov S et al (2008) A gene wiki for community annotation of gene function. PLoS Biol 6:e175
7. Menda N, Buels RM, Tecle I, Mueller LA (2008) A community-based annotation framework for linking solanaceae genomes with phenomes. Plant Physiol 147:1788–1799
8. Gene Ontology Consortium (2015) Gene Ontology Consortium: going forward. Nucleic Acids Res 43:D1049–1056
9. Leonelli S, Diehl AD, Christie KR, Harris MA, Lomax J (2011) How the gene ontology evolves. BMC Bioinformatics 12:325
10. Diehl AD, Lee JA, Scheuermann RH, Blake JA (2007) Ontology development for biological systems: immunology. Bioinformatics 23:913–915
11. Alam-Faruque Y, Hill DP, Dimmer EC, Harris MA, Foulger RE et al (2014) Representing kidney development using the gene ontology. PLoS One 9:e99864
12. Feltrin E, Campanaro S, Diehl AD, Ehler E, Faulkner G et al (2009) Muscle research and gene ontology: new standards for improved data integration. BMC Med Genomics 2:6
13. Tripathi S, Christie KR, Balakrishnan R, Huntley R, Hill DP et al (2013) Gene Ontology annotation of sequence-specific DNA binding transcription factors: setting the stage for a large-scale curation effort. Database (Oxford):bat062
14. Stein L (2001) Genome annotation: from sequence to biology. Nat Rev Genet 2:493–503
15. Camon E, Magrane M, Barrell D, Lee V, Dimmer E et al (2004) The Gene Ontology Annotation (GOA) Database: sharing knowledge in Uniprot with Gene Ontology. Nucleic Acids Res 32:D262–266
16. Balakrishnan R, Harris MA, Huntley R, Van Auken K, Cherry JM (2013) A guide to best practices for Gene Ontology (GO) manual annotation. Database (Oxford):bat054
17. Tweedie S, Ashburner M, Falls K, Leyland P, McQuilton P et al (2009) FlyBase: enhancing Drosophila Gene Ontology annotations. Nucleic Acids Res 37:D555–559
18. McDowall MD, Harris MA, Lock A, Rutherford K, Staines DM et al (2015) PomBase 2015: updates to the fission yeast database. Nucleic Acids Res 43:D656–661
19. Bradford Y, Conlin T, Dunn N, Fashena D, Frazer K et al (2011) ZFIN: enhancements and updates to the Zebrafish Model Organism Database. Nucleic Acids Res 39:D822–829
20. Huntley RP, Sawford T, Mutowo-Meullenet P, Shypitsyna A, Bonilla C et al (2015) The GOA database: gene Ontology annotation updates for 2015. Nucleic Acids Res 43: D1057–1063
21. Alam-Faruque Y, Dimmer EC, Huntley RP, O'Donovan C, Scambler P et al (2010) The Renal Gene Ontology Annotation Initiative. Organogenesis 6:71–75
22. Foulger RE, Denny P, Hardy J, Martin MJ, Sawford T, Lovering RC (2016) Using the gene ontology to annotate key players in Parkinson's disease. Neuroinformatics
23. Renfro DP, McIntosh BK, Venkatraman A, Siegele DA, Hu JC (2012) GONUTS: the Gene Ontology Normal Usage Tracking System. Nucleic Acids Res 40:D1262–1269
24. Patel S, Roncaglia P, Lovering RC (2015) Using Gene Ontology to describe the role of the neurexin-neuroligin-SHANK complex in human, mouse and rat and its relevance to autism. BMC Bioinformatics 16:186
25. Adams MD, Celniker SE, Holt RA, Evans CA, Gocayne JD et al (2000) The genome sequence of Drosophila melanogaster. Science 287:2185–2195
26. Brinkac L, Madupu R, Caler E, Harkins D, Lorenzi H, Thiagarajan M, Sutton G (2009) Expert assertions through community annotation Jamborees. Nature Precedings
27. van Dam TJ, Wheway G, Slaats GG, Group SS, Huynen MA et al (2013) The SYSCILIA gold standard (SCGSv1) of known ciliary components and its applications within a systems biology consortium. Cilia 2:7
28. Stables MJ, Shah S, Camon EB, Lovering RC, Newson J et al (2011) Transcriptomic analyses of murine resolution-phase macrophages. Blood 118:e192–208

7

Computational Methods for Annotation Transfers from Sequence

Domenico Cozzetto and David T. Jones

Abstract

Surveys of public sequence resources show that experimentally supported functional information is still completely missing for a considerable fraction of known proteins and is clearly incomplete for an even larger portion. Bioinformatics methods have long made use of very diverse data sources alone or in combination to predict protein function, with the understanding that different data types help elucidate complementary biological roles. This chapter focuses on methods accepting amino acid sequences as input and producing GO term assignments directly as outputs; the relevant biological and computational concepts are presented along with the advantages and limitations of individual approaches.

Key words Protein function prediction, Homology-based annotation transfers, Phylogenomics, Multi-domain architecture, De novo function prediction

1 Introduction

For decades experimentalists have been painstakingly probing a range of functional aspects of individual proteins. This steady but slow acquisition of functional data is in stark contrast to the results of next-generation sequencing technologies, which can survey gene expression regulation, genomic organization, and variation on a large scale [1]. Similarly, parallel efforts aim to map the networks of interactions between proteins, nucleic acids, and metabolites that regulate biological processes [2–4]. Nonetheless, comprehensive studies of protein function are hindered, because the combinations of gene products, biological roles, and cellular conditions are too numerous and because many experimental protocols cannot be applied to all proteins. Furthermore, the results need to be critically interpreted, integrated with existing knowledge, and translated into machine-readable formats—such as Gene Ontology (GO) [5] terms—for further analyses.

Manual curation requires substantial time and effort too; therefore the exponential growth in the number of sequences in UniProtKB [6] has only been matched by a linear increase in the number of entries with experimentally supported GO terms. Moreover, only 0.03 % of the sequences have received annotations for all three GO domains and the level of annotation detail can also fall far short of the maximum possible—e.g., there is direct evidence that some *E. coli K12* proteins act as transferases with no additional information about the chemical group relocated from the donor to the acceptor. Automated protein function prediction has consequently represented the only viable way to bridge some of these gaps, and indeed UniProtKB already exploits some computational tools (Fig. 1).

Given the lack of a general theory which can link protein sequences and environmental conditions directly to biological functions from physicochemical properties, current methods for protein function prediction implement knowledge-based heuristics that transfer functional information from already annotated proteins to unannotated ones. This chapter reviews sequence-based approaches to GO term prediction, which are the most popular, well understood, and easily accessible to a wide range of users. The

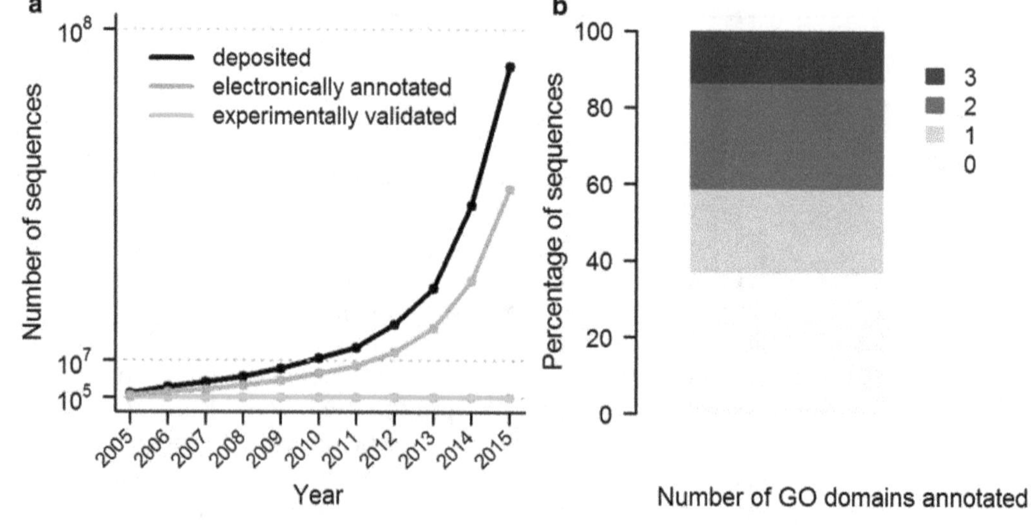

Fig. 1 Function annotation coverage of proteins in UniprotKB. (**a**) Over the past decade, the number of amino acid chains deposited in UniProtKB has grown exponentially (*black line*), while those with experimentally supported GO term assignments has only increased linearly (*green line*). This core subset however has allowed to assign GO terms to a substantial fraction of sequences (*orange line*). (**b**) Even with electronically inferred annotations, more than 80 % of sequences in UniProtKB release 2015_01 lack assignments for at least one of the molecular function, biological process, or cellular component GO sub-ontologies. Plots and statistics are based on the first release of each year

focus is primarily on the underpinning concepts and assumptions, as well as on the known advantages and pitfalls, which are all applicable to other controlled vocabularies, such as those described in the Chap. 19 [7] "KEGG, EC and other sources of functional data". How well current function prediction methods perform and how prediction accuracy can be measured are topics extensively covered in the Chap. 8 [8] "Evaluating GO annotations", Chap. 9 [9] "Evaluating functional annotations in enzymes", and Chap. 10 [10] "Community Assessment".

2 Annotation Transfers from Homologous Proteins

The most common way to annotate uncharacterized proteins consists in finding homologues—that is, proteins sharing common ancestry—of known function, and inheriting the information available for them under the assumption that function is evolutionarily conserved. BLAST [11] or PSI-BLAST [12] are routinely used to search for homologous sequences, and tools that compare sequences against hidden Markov models (HMMs), or pairs of profiles or of HMMs can be useful to extend the coverage of the protein sequence universe thanks to the increased sensitivity for remote homologues. A detailed presentation of sequence comparison methods is beyond the scope of this chapter and is available elsewhere [13]. In the simplest case, transfers can be made from the sequence with experimentally validated annotations and the lowest *E*-value—and this represents a useful baseline to benchmark the effectiveness of more advanced methods. This approach can produce erroneous results when key functional residues are mutated, or when the alignment doesn't span the whole length of the proteins—possibly indicating changes in domain architecture [14]. Iterative transfers of computationally generated functional assignments can lead to uncontrolled propagation of such errors; the average error rate of molecular function annotations is estimated to approach 0% only in the manually curated UniProtKB/SwissProt database, while it is substantially higher in un-reviewed resources [15].

Several studies have consequently attempted to estimate sequence similarity thresholds that would generate predictions with a guaranteed level of accuracy, and have suggested that 80% global sequence identity should be generally sufficient for safe annotation transfers [16–20]. However, this rule of thumb can either be too stringent or too lax, because biological sequences evolve at differing rates due to the need to maintain physiological function on the one hand, and to avoid deregulated gene expression, protein translation, folding, or physical interactions on the other [21]. Ideally, these cutoff values should be specific to individual families or even functional categories, but usually the number of labelled examples is not sufficient to allow reliable calibration. To circumvent these issues, it is possible to trade annotation specificity for accuracy, because broad functional aspects—e.g., about ligand binding and enzymatic or transporter activities—diverge at lower rates than the fine details—such as the specific metal ions bound or the molecules and chemical groups that are recognized and processed.

GOtcha [22] was the first tool to make predictions representing the enrichment of the GO terms assigned to BLAST hits in the hierarchical context of GO. It first calculates weights for each GO term, taking into account the number of similar sequences annotated with it and the statistical significance of the observed similarities. The program then considers the semantic relationships among the terms to update the tallies and reflect increasing confidence in more general annotations. PFP [23] follows a similar approach, but targets more difficult annotation cases, too, by leveraging information from PSI-BLAST hits with unconventionally high E-values. Furthermore, the scoring scheme exploits data about the co-occurrence of GO term pairs in UniProtKB entries, which allows safer annotations to be produced. Other methods fall in this category too, and interested readers are referred to the primary literature [24–27]. More sophisticated approaches rely on machine learning [28] rather than statistical analyses, and use experimental data to train classifiers that predict GO terms based on an array of alignment-derived features—such as sequence similarity scores, E-values, the coverage of the sequences, or the scores that GOtcha calculates for each GO category [29–31].

3 Annotation Transfers from Orthologous Proteins

Simple homology-based predictors are quick but error prone because they don't try to distinguish functionally equivalent relatives from those that have functionally diverged. In phylogenetic terms, this problem can be cast as classifying orthologues—homologue pairs evolved after speciation—and paralogues—homologue pairs derived from gene duplication. It is widely accepted that duplicated genes lack selective pressure to maintain their original biological roles, so they can easily undergo nucleotide changes ultimately leading to functional divergence [32]. The realization that genetic diversity arises from gene losses and horizontal transfers, too, makes phylogenetic reconstruction even more complex.

In this setup, annotations can be transferred with varying levels of confidence depending on how many orthologues there are and how closely related they are. This can partly account for the observation that orthologues can diverge functionally, particularly over long evolutionary distances or after duplication events in at least one of the lineages [33]. However, experimental studies have also shown that paralogues can retain functional equivalence, even long after the duplication event [34, 35]. Recent studies have consequently tested how useful the distinction between orthologues and paralogues is for protein function prediction and have drawn different conclusions [36–39]. The latest findings suggest that the functional similarity between orthologues is slightly higher than that between paralogues at the same level of sequence divergence, and that the signal is stronger for cellular components than for biological processes or molecular functions [38].

The traditional approach to orthologue detection involves computationally intensive calculations to build phylogenetic trees and then identify gene duplication and loss events [40]. SIFTER [41] builds on this framework to transfer the most specific experimentally supported molecular function terms available from the annotated sequences to all nodes in the tree using a Bayesian approach. The propagation algorithm captures the notion that functional transitions are more likely to occur after duplication than after speciation events, and when the terms are similar—i.e., the corresponding nodes are close in the GO graph. In order to speed up the computation, the authors have recently suggested limiting the number of GO term annotations that can be assigned to each protein [42], and they are providing pre-calculated predictions for a vast set of sequences from different species, including multi-domain proteins [43]. The semiautomated Phylogenetic Annotation and Inference Tool (PAINT) [44] recently adopted by the GO consortium provides a more flexible framework, which tries to keep functional change events uncoupled, so that the gain of one function does not imply the loss of another and vice versa—a desirable feature for annotating biological processes and for dealing with multifunctional proteins in general. Furthermore, unlike SIFTER, PAINT makes no assumption about how function diverges over evolutionary distance and whether its conservation is higher within orthologous groups than between them.

The increasing availability of completely sequenced genomes has promoted the development of alternative algorithms for orthologue detection. These first categorize pairs of orthologues in any two species, and then cluster the results across organisms, which helps recognize and fix spurious assignments [40]. The results are usually made publicly available in the form of specialized databases such as EggNOG [45], Ensembl Compara [46], Inparanoid [47], PANTHER [48], PhylomeDB [49], and OMA [50], and the clustering results provide the basis for GO term annotation transfers, under the assumption that the members of an orthologous group are functionally equivalent.

4 Annotation Transfers from Protein Families

Even when the sequence similarities between proteins of interest and those that have previously been characterized are limited to specific sites, such as individual domains or motifs, they can still be useful for function prediction. Some biological activities such as molecular recognition, protein targeting, and pathway regulation have long been mechanistically linked to short linear motifs—stretches of 10–20 consecutive amino acids exposed on protein surfaces [51]. Furthermore, some well-known protein families can be described by specific arrangements of multiple, possibly discontinuous, linear motifs, or by more general models of their domain sequences, namely sequence profiles [52] or hidden Markov models [53]. Many public databases now give access to groups of evolutionarily related proteins, coding for individual domains or multi-domain architectures. Even though these resources cannot directly assign GO terms to the input amino acid sequences, they can produce valuable assignments to know protein families.

InterPro [54] collates such results from 11 specialized and complementary resources, which differ by the types of patterns used for family assignment, by the amount of manual curation of their contents, and by the use of additional data such as 3-D structure or phylogenetic trees. InterPro entries combine available data and organize them in a hierarchical way, which mirrors the biological relationships between families and subfamilies of proteins. The curators also enrich these annotations with supporting biological information from the scientific literature and with links to external resources such as the PDB [55] and GO. InterPro provides function predictions for the input sequences based on the InterPro2GO mapping, which links each protein domain family to the most specific GO terms that apply to all its members [56]. These annotations form a large bulk of the electronically inferred functional assignments in UniProtKB, where they are integrated with associations generated from other controlled vocabularies, e.g., about subcellular localization and enzymatic activity.

CATH-Gene3D [57] and SUPERFAMILY [58] are two databases that store domain assignments for known protein sequences based on the CATH [59] and SCOP [60] protein structure classification schemes, respectively. CATH-Gene3D data are clustered into functional families which include relatives with highly similar sequences, structures, and functions, as to highlight the strong conservation of important regions such as specificity-determining residues. GO terms are associated probabilistically to each functional family based on how often they occur in the UniProtKB annotations of the whole sequences. The recent CATH FunFHMMer web server automates the search procedure for input sequences, resolves multi-domain architectures, assigns each predicted domain to its functional family, and finally inherits the GO term annotations found in the library [61]. The dcGO—short for domain centric—method follows a similar route, but with some key differences [62]. HMM models are built for both individual domains and supra-domains, i.e., sets of consecutive domains that are defined according to the SCOP structural definition and the evolutionary one in Pfam [63]. Given the annotations in the GOA database [64] and the GO hierarchical structure, each domain and supra-domain is labelled with a set of GO terms that are associated with it in a statistically significant way. The strength of each association is then empirically converted into a confidence score. To facilitate the analysis of the results by non-specialists, the predicted GO terms are divided into four classes according to how specific and informative they are using their information content.

5 De Novo Function Annotation Using Biological Features

The function annotation methods described so far make use of homology to transfer GO terms to a target protein from other previously characterized proteins. In some cases, however, no useful functional annotations can be found for any of the detectable homologues, or in the most extreme case no homologous sequences can be found at all. In this case a de novo method is required which can infer GO terms directly from amino acid sequence in the absence of evolutionary relatedness. This is a very hard problem, and only a few tools have been developed which can handle these situations. The most successful approaches to date employ the basic idea of first transforming the target sequence into a set of component features. These features are then related to particular broad functional classes by means of supervised machine learning techniques. In this way the methods address the question of what kinds of functions can proteins perform with the given set of protein features. As a trivial example, proteins which are predicted to have particular numbers of transmembrane helices as component features will be more likely to have transmembrane transporter activity.

ProtFun, which makes use of neural networks, was the first widely used method for transferring functional annotations between human proteins through similarity of biochemical attributes, such as the occurrence of charged amino acids, low-complexity regions, signal peptides, trans-membrane helices, and posttranslationally modified residues [65, 66]. In the original ProtFun method, only the broad functional classes originally compiled by Monica Riley [67] were considered, but later the authors extended their approach to predicting a representative set of GO terms. FFPred, which is based on support vector machines, has taken this approach further by considering the observed strong correlation between the lengths and positions of intrinsically disordered protein regions with certain molecular functions and biological processes [68, 69]. As with

ProtFun, FFPred was initially developed specifically for annotating human proteins, but the results have been shown to extend reasonably well to other vertebrate proteomes too.

Feature-based protein function assignment offers both advantages and disadvantages over sequence similarity-based approaches. The main advantage is fairly obvious: feature-based methods can work in the absence of homology to characterized proteins, and thus can even be used to assign GO terms to orphan proteins. A further advantage is that feature-based prediction is also able to provide insight into functional changes that occur after alternative splicing, as the input features are likely to reflect sequence deletions relative to the main transcript, e.g., the loss of a signal peptide or disordered region. Probably the main disadvantage is that classification models can only be built for GO terms where there are sufficient examples with experimentally validated assignments. This generally means that assignments can only be made for terms fairly high up in the overall GO graph, and thus highly specific predictions are generally not possible using this kind of approach. Of course, as datasets become larger, these methods will be able to overcome such limitation.

6 Conclusions and Outlook

The widening gap between the number of known sequences and those experimentally characterized has stimulated the development and refinement of a wide array of computational methods for protein function prediction. The scope of this survey has been limited to four classes of sequence-based approaches for GO term annotation transfers, but several other routes could be followed. If the 3-D structure of a protein has been solved or accurately modelled, it is possible to search for global or local structural similarities and predict binding regions and catalytic sites [70, 71]. Comparison of multiple complete genomes can help detect not only orthologous genes as described above, but also further patterns indicative of functional linkages between gene pairs such as fusion events, conserved chromosomal proximity, and co-occurrence/absence in a group of species [72]. Phylogenetic profiling posits that coevolving protein families are functionally coupled, e.g., because they encode for proteins assembling into obligate complexes or participating in the same biological process. Since its inception, this "guilt-by-association" method has been implemented in several different ways [73], and tools able to make GO term assignments are also emerging [74]. Involvement in the same biological process or co-localization can also be inferred from the analysis of protein-protein interaction maps, gene expression profiles, and phenotypic variations following engineered genetic mutations [75]. Finally, integrative strategies combine all such heterogeneous data sources and hold the potential to produce more confident predictions, reduce errors, and overcome the intrinsic limitations of individual algorithms [31, 76–78]. For instance, protein sequence and structure data appear to be better suited to predict terms in the molecular function category, while genome-wide datasets can shed light on biological processes and protein subcellular localization. In the future, these methods will become increasingly valuable to generate testable hypotheses about protein function as they improve in accuracy – thanks to additional experimental data and to better ways of using them – as well as in user-friendliness to experimentalists and nonspecialists in general.

Acknowledgements

This work was partially supported by the UK Biotechnology and Biological Sciences Research Council. Open Access charges were funded by the University College London Library, the Swiss Institute of Bioinformatics, the Agassiz Foundation, and the Foundation for the University of Lausanne.

References

1. Soon WW, Hariharan M, Snyder MP (2013) High-throughput sequencing for biology and medicine. Mol Syst Biol 9:640. doi:10.1038/msb.2012.61
2. Mitra K, Carvunis AR, Ramesh SK, Ideker T (2013) Integrative approaches for finding modular structure in biological networks. Nat Rev Genet 14(10):719–732. doi:10.1038/nrg3552
3. Mahony S, Pugh BF (2015) Protein-DNA binding in high-resolution. Crit Rev Biochem Mol Biol:1–15. doi:10.3109/10409238.2015.1051505
4. McHugh CA, Russell P, Guttman M (2014) Methods for comprehensive experimental identification of RNA-protein interactions. Genome Biol 15(1):203. doi:10.1186/gb4152
5. Ashburner M, Ball CA, Blake JA, Botstein D, Butler H, Cherry JM, Davis AP, Dolinski K, Dwight SS, Eppig JT, Harris MA, Hill DP, Issel-Tarver L, Kasarskis A, Lewis S, Matese JC, Richardson JE, Ringwald M, Rubin GM, Sherlock G (2000) Gene ontology: tool for the unification of biology. The Gene Ontology Consortium. Nat Genet 25(1):25–29. doi:10.1038/75556
6. UniProt C (2015) UniProt: a hub for protein information. Nucleic Acids Res 43(Database issue):D204–D212. doi:10.1093/nar/gku989
7. Furnham N (2016) Complementary sources of protein functional information: the far side of GO. In: Dessimoz C, Škunca N (eds) The gene ontology handbook. Methods in molecular biology, vol 1446. Humana Press. Chapter 19
8. Škunca N, Roberts RJ, Steffen M (2016) Evaluating computational gene ontology annotations. In: Dessimoz C, Škunca N (eds) The gene ontology handbook. Methods in molecular biology, vol 1446. Humana Press. Chapter 8
9. Holliday GL, Davidson R, Akiva E, Babbitt PC (2016) Evaluating functional annotations of enzymes using the gene ontology. In: Dessimoz C, Škunca N (eds) The gene ontology handbook. Methods in molecular biology, vol 1446. Humana Press. Chapter 9
10. Friedberg I, Radivojac P (2016) Community-wide evaluation of computational function prediction. In: Dessimoz C, Škunca N (eds) The gene ontology handbook. Methods in molecular biology, vol 1446. Humana Press. Chapter 10
11. Altschul SF, Gish W, Miller W, Myers EW, Lipman DJ (1990) Basic local alignment search tool. J Mol Biol 215(3):403–410. doi:10.1016/S0022-2836(05)80360-2

12. Altschul SF, Madden TL, Schaffer AA, Zhang J, Zhang Z, Miller W, Lipman DJ (1997) Gapped BLAST and PSI-BLAST: a new generation of protein database search programs. Nucleic Acids Res 25(17):3389–3402
13. Soding J, Remmert M (2011) Protein sequence comparison and fold recognition: progress and good-practice benchmarking. Curr Opin Struct Biol 21(3):404–411. doi:10.1016/j.sbi.2011.03.005
14. Rost B (2002) Enzyme function less conserved than anticipated. J Mol Biol 318(2):595–608. doi:10.1016/S0022-2836(02)00016-5
15. Schnoes AM, Brown SD, Dodevski I, Babbitt PC (2009) Annotation error in public databases: misannotation of molecular function in enzyme superfamilies. PLoS Comput Biol 5(12): e1000605. doi:10.1371/journal.pcbi.1000605
16. Devos D, Valencia A (2000) Practical limits of function prediction. Proteins 41(1):98–107
17. Wilson CA, Kreychman J, Gerstein M (2000) Assessing annotation transfer for genomics: quantifying the relations between protein sequence, structure and function through traditional and probabilistic scores. J Mol Biol 297(1):233–249.doi:10.1006/jmbi.2000.3550
18. Tian W, Skolnick J (2003) How well is enzyme function conserved as a function of pairwise sequence identity? J Mol Biol 333(4):863–882
19. Sangar V, Blankenberg DJ, Altman N, Lesk AM (2007) Quantitative sequence-function relationships in proteins based on gene ontology. BMC Bioinformatics 8:294. doi:10.1186/1471-2105-8-294
20. Addou S, Rentzsch R, Lee D, Orengo CA (2009) Domain-based and family-specific sequence identity thresholds increase the levels of reliable protein function transfer. J Mol Biol 387(2):416–430. doi:10.1016/j.jmb.2008.12.045
21. Zhang J, Yang JR (2015) Determinants of the rate of protein sequence evolution. Nat Rev Genet 16(7):409–420. doi:10.1038/nrg3950
22. Martin DM, Berriman M, Barton GJ (2004) GOtcha: a new method for prediction of protein function assessed by the annotation of seven genomes. BMC Bioinformatics 5:178. doi:10.1186/1471-2105-5-178
23. Hawkins T, Chitale M, Luban S, Kihara D (2009) PFP: automated prediction of gene ontology functional annotations with confidence scores using protein sequence data. Proteins 74(3):566–582. doi:10.1002/prot.22172
24. Chitale M, Hawkins T, Park C, Kihara D (2009) ESG: extended similarity group method for automated protein function prediction. Bioinformatics 25(14):1739–1745. doi:10.1093/bioinformatics/btp309
25. Vinayagam A, Konig R, Moormann J, Schubert F, Eils R, Glatting KH, Suhai S (2004) Applying support vector machines for gene ontology based gene function prediction. BMC Bioinformatics 5:116. doi:10.1186/1471-2105-5-116
26. Gotz S, Garcia-Gomez JM, Terol J, Williams TD, Nagaraj SH, Nueda MJ, Robles M, Talon M, Dopazo J, Conesa A (2008) High-throughput functional annotation and data mining with the Blast2GO suite. Nucleic Acids Res 36(10):3420–3435. doi:10.1093/nar/gkn176
27. Piovesan D, Martelli PL, Fariselli P, Zauli A, Rossi I, Casadio R (2011) BAR-PLUS: the Bologna Annotation Resource Plus for functional and structural annotation of protein sequences. Nucleic Acids Res 39(Web Server issue):W197–W202. doi:10.1093/nar/gkr292
28. Duda RO, Hart PE, Stork DG (2012) Pattern classification. Wiley, New York
29. Sokolov A, Ben-Hur A (2010) Hierarchical classification of gene ontology terms using the GOstruct method. J Bioinforma Comput Biol 8(02):357–376
30. Clark WT, Radivojac P (2011) Analysis of protein function and its prediction from amino acid sequence. Proteins 79(7):2086–2096
31. Cozzetto D, Buchan DW, Bryson K, Jones DT (2013) Protein function prediction by massive integration of evolutionary analyses and multiple data sources. BMC Bioinformatics 14(Suppl 3):S1. doi:10.1186/1471-2105-14-S3-S1
32. Gabaldon T, Koonin EV (2013) Functional and evolutionary implications of gene orthology. Nat Rev Genet 14(5):360–366. doi:10.1038/nrg3456
33. Kachroo AH, Laurent JM, Yellman CM, Meyer AG, Wilke CO, Marcotte EM (2015) Evolution. Systematic humanization of yeast genes reveals conserved functions and genetic modularity. Science 348(6237):921–925. doi:10.1126/science.aaa0769
34. Dean EJ, Davis JC, Davis RW, Petrov DA (2008) Pervasive and persistent redundancy among duplicated genes in yeast. PLoS Genet 4(7): e1000113. doi:10.1371/journal.pgen.1000113
35. Tischler J, Lehner B, Chen N, Fraser AG (2006) Combinatorial RNA interference in Caenorhabditis elegans reveals that redundancy between gene duplicates can be maintained for more than 80 million years of evolution. Genome Biol 7(8):R69. doi:10.1186/gb-2006-7-8-R69

36. Nehrt NL, Clark WT, Radivojac P, Hahn MW (2011) Testing the ortholog conjecture with comparative functional genomic data from mammals. PLoS Comput Biol 7(6):e1002073. doi:10.1371/journal.pcbi.1002073
37. Chen X, Zhang J (2012) The ortholog conjecture is untestable by the current gene ontology but is supported by RNA sequencing data. PLoS Comput Biol 8(11):e1002784. doi:10.1371/journal.pcbi.1002784
38. Altenhoff AM, Studer RA, Robinson-Rechavi M, Dessimoz C (2012) Resolving the ortholog conjecture: orthologs tend to be weakly, but significantly, more similar in function than paralogs. PLoS Comput Biol 8(5):e1002514. doi:10.1371/journal.pcbi.1002514
39. Rogozin IB, Managadze D, Shabalina SA, Koonin EV (2014) Gene family level comparative analysis of gene expression in mammals validates the ortholog conjecture. Genome Biol Evol 6(4):754–762. doi:10.1093/gbe/evu051
40. Altenhoff AM, Dessimoz C (2012) Inferring orthology and paralogy. Methods Mol Biol 855:259–279. doi:10.1007/978-1-61779-582-4_9
41. Engelhardt BE, Jordan MI, Muratore KE, Brenner SE (2005) Protein molecular function prediction by Bayesian phylogenomics. PLoS Comput Biol 1(5):e45. doi:10.1371/journal.pcbi.0010045
42. Engelhardt BE, Jordan MI, Srouji JR, Brenner SE (2011) Genome-scale phylogenetic function annotation of large and diverse protein families. Genome Res 21(11):1969–1980. doi:10.1101/gr.104687.109
43. Sahraeian SM, Luo KR, Brenner SE (2015) SIFTER search: a web server for accurate phylogeny-based protein function prediction. Nucleic Acids Res. doi:10.1093/nar/gkv461
44. Gaudet P, Livstone MS, Lewis SE, Thomas PD (2011) Phylogenetic-based propagation of functional annotations within the Gene Ontology consortium. Brief Bioinform 12(5):449–462. doi:10.1093/bib/bbr042
45. Huerta-Cepas J, Szklarczyk D, Forslund K, Cook H, Heller D, Walter MC, Rattei T, Mende DR, Sunagawa S, Kuhn M, Jensen LJ, von Mering C, Bork P (2016) eggNOG 4.5: a hierarchical orthology framework with improved functional annotations for eukaryotic, prokaryotic and viral sequences. Nucleic Acids Res 44(D1):D286–D293. doi:10.1093/nar/gkv1248
46. Flicek P, Amode MR, Barrell D, Beal K, Billis K, Brent S, Carvalho-Silva D, Clapham P, Coates G, Fitzgerald S, Gil L, Giron CG, Gordon L, Hourlier T, Hunt S, Johnson N, Juettemann T, Kahari AK, Keenan S, Kulesha E, Martin FJ, Maurel T, McLaren WM, Murphy DN, Nag R, Overduin B, Pignatelli M, Pritchard B, Pritchard E, Riat HS, Ruffier M, Sheppard D, Taylor K, Thormann A, Trevanion SJ, Vullo A, Wilder SP, Wilson M, Zadissa A, Aken BL, Birney E, Cunningham F, Harrow J, Herrero J, Hubbard TJ, Kinsella R, Muffato M, Parker A, Spudich G, Yates A, Zerbino DR, Searle SM (2014) Ensembl 2014. Nucleic Acids Res 42(Database issue):D749–D755. doi:10.1093/nar/gkt1196
47. Sonnhammer EL, Ostlund G (2015) InParanoid 8: orthology analysis between 273 proteomes, mostly eukaryotic. Nucleic Acids Res 43(Database issue):D234–D239. doi:10.1093/nar/gku1203
48. Mi H, Poudel S, Muruganujan A, Casagrande JT, Thomas PD (2016) PANTHER version 10: expanded protein families and functions, and analysis tools. Nucleic Acids Res 44(D1): D336–D342. doi:10.1093/nar/gkv1194
49. Huerta-Cepas J, Capella-Gutierrez S, Pryszcz LP, Marcet-Houben M, Gabaldon T (2014) PhylomeDB v4: zooming into the plurality of evolutionary histories of a genome. Nucleic Acids Res 42(Database issue):D897–D902. doi:10.1093/nar/gkt1177
50. Altenhoff AM, Skunca N, Glover N, Train CM, Sueki A, Pilizota I, Gori K, Tomiczek B, Muller S, Redestig H, Gonnet GH, Dessimoz C (2015) The OMA orthology database in 2015: function predictions, better plant support, synteny view and other improvements. Nucleic Acids Res 43(Database issue):D240–D249. doi:10.1093/nar/gku1158
51. Van Roey K, Uyar B, Weatheritt RJ, Dinkel H, Seiler M, Budd A, Gibson TJ, Davey NE (2014) Short linear motifs: ubiquitous and functionally diverse protein interaction modules directing cell regulation. Chem Rev 114(13):6733–6778. doi:10.1021/cr400585q
52. Gribskov M, McLachlan AD, Eisenberg D (1987) Profile analysis: detection of distantly related proteins. Proc Natl Acad Sci U S A 84(13):4355–4358
53. Eddy SR (1998) Profile hidden Markov models. Bioinformatics 14(9):755–763
54. Mitchell A, Chang HY, Daugherty L, Fraser M, Hunter S, Lopez R, McAnulla C, McMenamin C, Nuka G, Pesseat S, Sangrador-Vegas A, Scheremetjew M, Rato C, Yong SY, Bateman A, Punta M, Attwood TK, Sigrist CJ, Redaschi N, Rivoire C, Xenarios I, Kahn D, Guyot D, Bork P, Letunic I, Gough J, Oates M, Haft D, Huang H, Natale DA, Wu CH, Orengo C,

Sillitoe I, Mi H, Thomas PD, Finn RD (2015) The InterPro protein families database: the classification resource after 15 years. Nucleic Acids Res 43(Database issue):D213–D221. doi:10.1093/nar/gku1243

55. Berman HM, Westbrook J, Feng Z, Gilliland G, Bhat TN, Weissig H, Shindyalov IN, Bourne PE (2000) The Protein Data Bank. Nucleic Acids Res 28(1):235–242
56. Burge S, Kelly E, Lonsdale D, Mutowo-Muellenet P, McAnulla C, Mitchell A, Sangrador-Vegas A, Yong SY, Mulder N, Hunter S (2012) Manual GO annotation of predictive protein signatures: the InterPro approach to GO curation. Database (Oxford) 2012:bar068. doi:10.1093/database/bar068
57. Sillitoe I, Lewis TE, Cuff A, Das S, Ashford P, Dawson NL, Furnham N, Laskowski RA, Lee D, Lees JG, Lehtinen S, Studer RA, Thornton J, Orengo CA (2015) CATH: comprehensive structural and functional annotations for genome sequences. Nucleic Acids Res 43(Database issue):D376–D381. doi:10.1093/nar/gku947
58. Oates ME, Stahlhacke J, Vavoulis DV, Smithers B, Rackham OJ, Sardar AJ, Zaucha J, Thurlby N, Fang H, Gough J (2015) The SUPERFAMILY 1.75 database in 2014: a doubling of data. Nucleic Acids Res 43(Database issue):D227–D233. doi:10.1093/nar/gku1041
59. Orengo CA, Michie AD, Jones S, Jones DT, Swindells MB, Thornton JM (1997) CATH--a hierarchic classification of protein domain structures. Structure 5(8):1093–1108
60. Murzin AG, Brenner SE, Hubbard T, Chothia C (1995) SCOP: a structural classification of proteins database for the investigation of sequences and structures. J Mol Biol 247(4): 536–540. doi:10.1006/jmbi.1995.0159
61. Das S, Sillitoe I, Lee D, Lees JG, Dawson NL, Ward J, Orengo CA (2015) CATH FunFHMMer web server: protein functional annotations using functional family assignments. Nucleic Acids Res. doi:10.1093/nar/gkv488
62. Fang H, Gough J (2013) DcGO: database of domain-centric ontologies on functions, phenotypes, diseases and more. Nucleic Acids Res 41(Database issue):D536–D544. doi:10.1093/nar/gks1080
63. Finn RD, Bateman A, Clements J, Coggill P, Eberhardt RY, Eddy SR, Heger A, Hetherington K, Holm L, Mistry J, Sonnhammer EL, Tate J, Punta M (2014) Pfam: the protein families database. Nucleic Acids Res 42(Database issue):D222–D230. doi:10.1093/nar/gkt1223
64. Huntley RP, Sawford T, Mutowo-Meullenet P, Shypitsyna A, Bonilla C, Martin MJ, O'Donovan C (2015) The GOA database: gene Ontology annotation updates for 2015. Nucleic Acids Res 43(Database issue):D1057–D1063. doi:10.1093/nar/gku1113
65. Jensen LJ, Gupta R, Blom N, Devos D, Tamames J, Kesmir C, Nielsen H, Staerfeldt HH, Rapacki K, Workman C, Andersen CA, Knudsen S, Krogh A, Valencia A, Brunak S (2002) Prediction of human protein function from post-translational modifications and localization features. J Mol Biol 319(5):1257–1265. doi:10.1016/S0022-2836(02)00379-0
66. Jensen LJ, Gupta R, Staerfeldt HH, Brunak S (2003) Prediction of human protein function according to Gene Ontology categories. Bioinformatics 19(5):635–642
67. Riley M (1993) Functions of the gene products of Escherichia coli. Microbiol Rev 57(4):862–952
68. Lobley A, Swindells MB, Orengo CA, Jones DT (2007) Inferring function using patterns of native disorder in proteins. PLoS Comput Biol 3(8):e162. doi:10.1371/journal.pcbi.0030162
69. Minneci F, Piovesan D, Cozzetto D, Jones DT (2013) FFPred 2.0: improved homology-independent prediction of gene ontology terms for eukaryotic protein sequences. PLoS One 8(5):e63754. doi:10.1371/journal.pone.0063754
70. Jacobson MP, Kalyanaraman C, Zhao S, Tian B (2014) Leveraging structure for enzyme function prediction: methods, opportunities, and challenges. Trends Biochem Sci 39(8):363–371. doi:10.1016/j.tibs.2014.05.006
71. Petrey D, Chen TS, Deng L, Garzon JI, Hwang H, Lasso G, Lee H, Silkov A, Honig B (2015) Template-based prediction of protein function. Curr Opin Struct Biol 32C:33–38. doi:10.1016/j.sbi.2015.01.007
72. Galperin MY, Koonin EV (2014) Comparative genomics approaches to identifying functionally related genes. In: Algorithms for computational biology. Springer, Berlin, pp 1–24
73. Pellegrini M (2012) Using phylogenetic profiles to predict functional relationships. Methods Mol Biol 804:167–177. doi:10.1007/978-1-61779-361-5_9
74. Skunca N, Bosnjak M, Krisko A, Panov P, Dzeroski S, Smuc T, Supek F (2013) Phyletic

profiling with cliques of orthologs is enhanced by signatures of paralogy relationships. PLoS Comput Biol 9(1):e1002852. doi:10.1371/journal.pcbi.1002852

75. Yu D, Kim M, Xiao G, Hwang TH (2013) Review of biological network data and its applications. Genomics Inform 11(4):200–210. doi:10.5808/GI.2013.11.4.200
76. Ma X, Chen T, Sun F (2014) Integrative approaches for predicting protein function and prioritizing genes for complex phenotypes using protein interaction networks. Brief Bioinform 15(5):685–698. doi:10.1093/bib/bbt041
77. Wass MN, Barton G, Sternberg MJ (2012) CombFunc: predicting protein function using heterogeneous data sources. Nucleic Acids Res 40(Web Server issue):W466–W470. doi:10.1093/nar/gks489
78. Piovesan D, Giollo M, Leonardi E, Ferrari C, Tosatto SC (2015) INGA: protein function prediction combining interaction networks, domain assignments and sequence similarity. Nucleic Acids Res. doi:10.1093/nar/gkv523

Evaluating Gene Ontology Annotations

8

Evaluating Functional Annotations of Enzymes using the Gene Ontology

Gemma L. Holliday, Rebecca Davidson, Eyal Akiva, and Patricia C. Babbitt

Abstract

The Gene Ontology (GO) (Ashburner et al., Nat Genet 25(1):25–29, 2000) is a powerful tool in the informatics arsenal of methods for evaluating annotations in a protein dataset. From identifying the nearest well annotated homologue of a protein of interest to predicting where misannotation has occurred to knowing how confident you can be in the annotations assigned to those proteins is critical. In this chapter we explore what makes an enzyme unique and how we can use GO to infer aspects of protein function based on sequence similarity. These can range from identification of misannotation or other errors in a predicted function to accurate function prediction for an enzyme of entirely unknown function. Although GO annotation applies to any gene products, we focus here a describing our approach for hierarchical classification of enzymes in the Structure-Function Linkage Database (SFLD) (Akiva et al., Nucleic Acids Res 42(Database issue):D521–530, 2014) as a guide for informed utilisation of annotation transfer based on GO terms.

Key words Catalytic function, Enzyme, Misannotation, Evidence of function

1 Introduction

Enzymes are the biological toolkit that organisms use to perform the chemistry of life, and the Gene Ontology (GO) [1] represents a detailed vocabulary of annotations that captures many of the functional nuances of these proteins. However, the relative lack of experimentally validated annotations means that the vast majority of functional annotations are electronically transferred, which can lead to erroneous assumptions and missannotations. Thus, it is important to be able to critically examine functional annotations. This chapter describes some of the key concepts that are unique for applying GO-assisted annotation to enzymes. In particular we introduce several techniques to assess their functional annotation within the framework of evolutionarily related proteins (superfamilies).

1.1 Enzyme Nomenclature and How It Is Used in GO

At its very simplest, an enzyme is a protein that can perform at least one overall chemical transformation (the function of the enzyme). The overall chemical transformation is often described by the Enzyme Commission (EC) Number [2–4] (and *see* Chap. 19 [5]). The EC Number takes the form A.B.C.D, where each position in the code is a number. The first number (which ranges from 1 to 6) describes the general class of enzyme, the second two numbers (which both range from 1 to 99) describe the chemical changes occurring in more detail (the exact meaning of the numbers depends on the specific class of enzyme you are looking at) and the final number (formally ranging from 1 to 999) essentially describes the substrate specificity. The EC number has many limitations, not least the fact that it doesn't describe the mechanism (the manner in which the enzyme performs its overall reaction) and often contains no information on cofactors, regulators, etc. Nor is it structurally contextual [6] in that similarity in EC number does not necessarily infer similarity in sequence or structure, making it sometimes risky to use for annotation transfer, especially among remote homologous proteins. However, it does do exactly what it says on the tin: it defines the overall chemical transformation. This makes it an important and powerful tool for many applications that require a description of enzyme chemistry.

The Molecular Function Ontology (MFO) in GO contains the full definition of around 70% of all currently available EC numbers. Theoretically, the MFO would contain all EC numbers available. However, due to many EC numbers not currently being assigned to a specific protein identifier within UniProtKB, the coverage is lower than might be expected. Another important difference between the EC hierarchy and the GO hierarchy is that the latter is often much more complex than the simple four steps found in the EC hierarchy. For example, the biotin synthase (EC 2.8.1.6) hierarchy is relatively simple and follows the four step nomenclature, while the GO hierarchy for [cytochrome c]-arginine N-methyltransferase (EC 2.1.1.124) is much more complex (*see* Fig. 1).

Formally, MFO terms describe the activities that occur at the molecular level; this includes the "catalytic activity" of enzymes or "binding activity". It is important to remember that EC numbers and MFO terms represent activities and not the entities (molecules, proteins or complexes) that perform them. Further, they do not specify where, when or in what context the action takes place. This is usually handled by the Cellular Component Ontology. The final ontology in GO, the Biological Process Ontology (BPO), provides terms to describe a series of events that are accomplished by one or more organised assemblies of molecular functions. Each MFO term describes a unique single function that means the same thing regardless of the evolutionary origin of the entity annotated with that term. Although the BPO describes a collection of activities, some BPO terms can be related

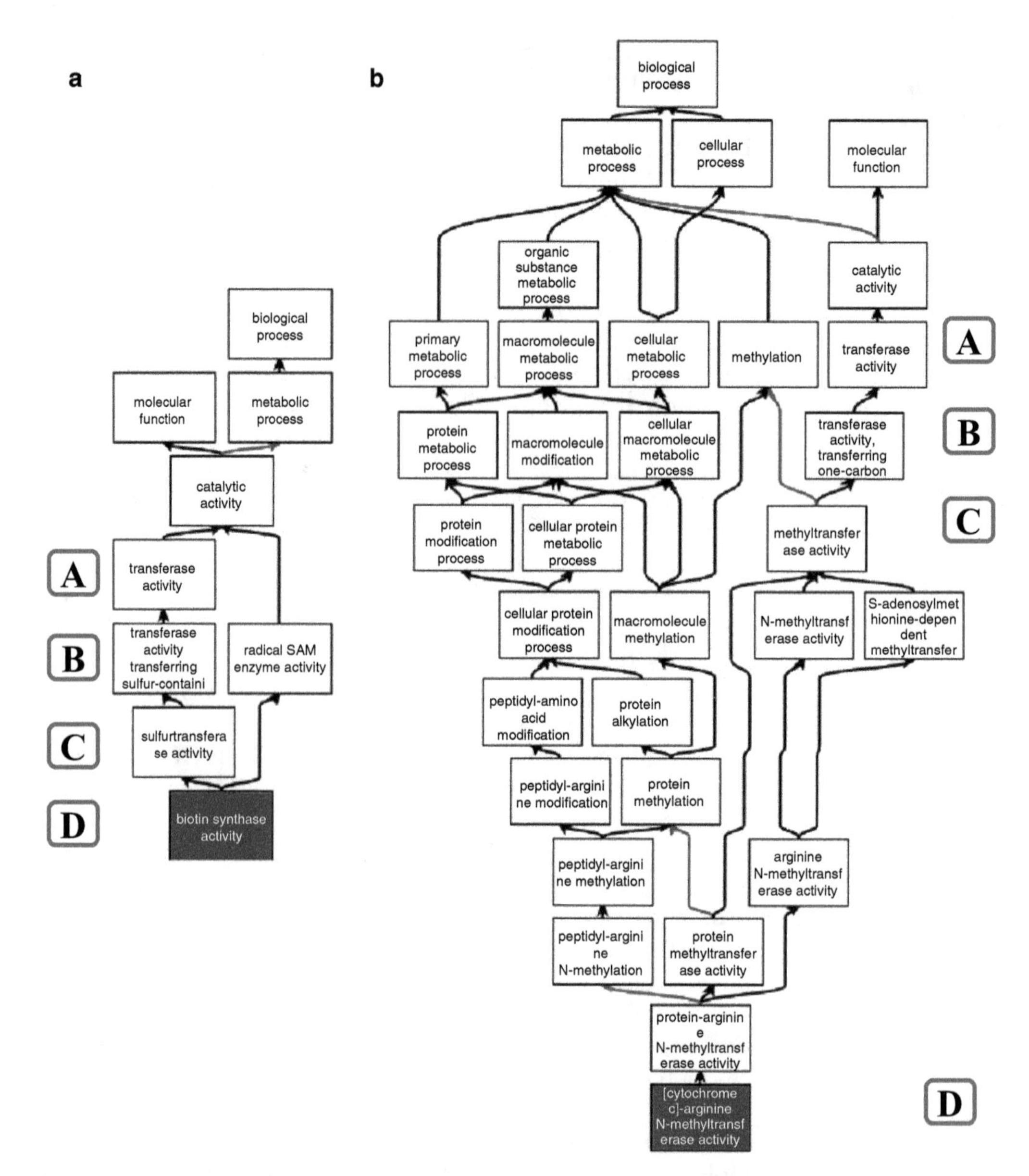

Fig. 1 Example of the GO hierarchy (taken from the ancestor chart of the QuickGo website (http://www.ebi.ac.uk/QuickGO/) showing the relative complexity of the GO hierarchy for two distinct EC numbers). (**a**) Shows the GO hierarchy for biotin synthase, EC 2.8.1.6; (**b**) shows the GO hierarchy for [cytochrome c]-arginine *N*-methyltransferase, EC 2.1.1.24. The *colours* of the *arrows* in the ontology are denoted by the key in the centre of the figure. *Black connections* between terms represent an is_a relationship, *blue connections* represent a part_of relationship. The *A*, *B*, *C* and *D* in *red boxes* denote the four levels of the EC nomenclature

to their counterparts in the MFO, e.g. GO:0009102 (biotin biosynthetic process) could be considered to be subsumed with the MFO term GO:0004076 (biotin synthase activity) as GO:0009102 includes the activity GO:0004076, i.e. in such cases, the terms are interchangeable for the purpose of evaluation of a protein's

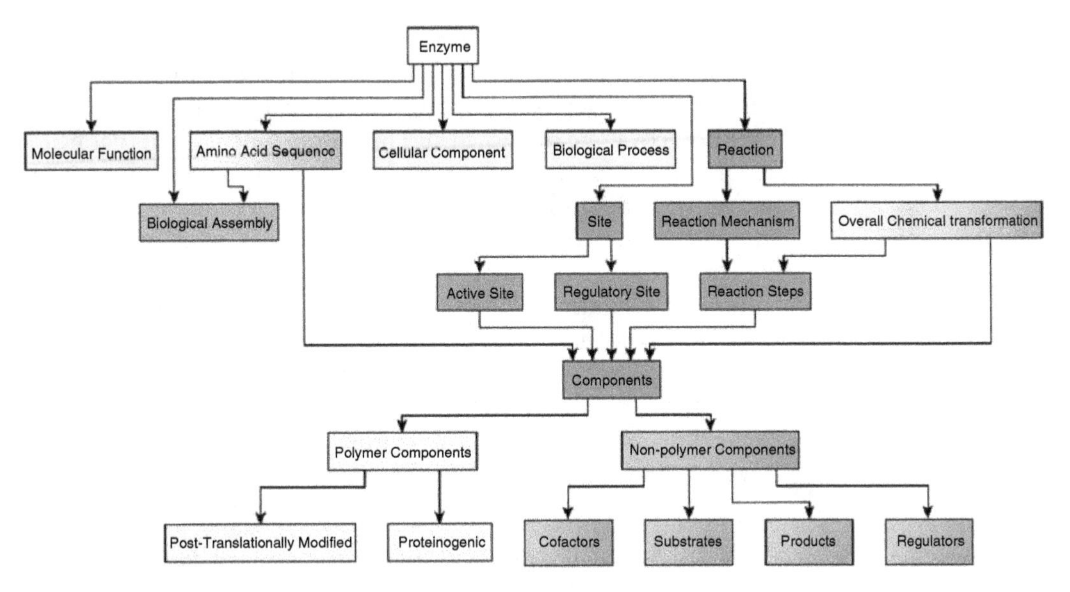

Fig. 2 Hierarchical view of enzyme features. The GO ontologies which describe proteins and their features are highlighted in *light green*. Other ontologies available in OBO and BioPortal are shown in the following colours: *light yellow* represents the Amino Acid Ontology, *purple* represents the Enzyme Mechanism Ontology, *blue* represents the ChEBI ontology and *grey* represents the Protein Ontology. *See* also Chap. 5 [10]. The terms immediately beneath the parent term are those terms that are covered by ontologies, and required for a protein to be considered an enzyme

annotation. Please *see* Chap. 2 [7] for a more in-depth discussion of the differences between BPO and MFO.

As a protein, an enzyme has many features that can be described and used to define the enzyme's function, from the primary amino acid sequence to the enzyme's quaternary structure (biological assembly), the chemistry that is catalysed, to the localisation of the enzyme. Features can also denote the presence (or absence) of active site residues to confirm (or deny) a predicted function, such as EC class, using the compositional makeup of a protein amino acid sequences [8, 9]. Nevertheless, for the many proteins of unknown function deposited in genome projects, prediction of the molecular, biological, and cellular functions remains a daunting challenge. Figure 2 provides a view of enzyme-specific features along with the GO ontologies that can also be used to describe them. Because it captures these features through a systematic and hierarchical classification system, GO is heavily used as a standard for evaluation of function prediction methods. For example, a regular competition, the Critical Assessment of Functional Annotation (CAFA) has brought many in the function prediction community together to evaluate automated protein function prediction algorithms in assigning GO terms to protein sequences [11]. Please *see* Chap. 10 [12] for a more detailed discussion of CAFA.

1.2 Why Annotate Enzymes with the Gene Ontology?

Although there are many different features and methods that can (and are) used to predict the function of a protein, there are several advantages to using GO as a broadly applied standard. Firstly, GO has good coverage of known and predicted functions so that nearly all proteins in GO will have at least one associated annotation. Secondly, annotations associated with a protein are accompanied by an evidence code, along with the information describing that evidence source. Within the SFLD [13] each annotation has an associated confidence level which is linked to both the evidence code, source of the evidence (including the type of experiment) and the curator's experience. For example, experimental evidence for an annotation is considered as having high confidence whereas predictions generated by computational methods are considered of lower confidence (Chap. 3 [14]). In general there are three types of evidence for the assignment of a GO term to a protein:

1. Fully manually curated: These proteins will usually have an associated experimental evidence that has been identified by human curators and who have added relevant evidence codes. For the purposes of the SFLD and this chapter, these are considered high confidence and will have a greater weight than any other annotation confidence level.
2. Computational with some curator input: These are computationally based annotations that have been propagated through curator derived rules, and are generally considered to be of medium confidence by the SFLD. Due to the huge proportion of sequences in large public databases now available, over 98 % of GO annotations are inferred computationally [15].
3. Computational with no curator input: These annotations that have been computationally inferred from information without any curator input into the inference rules and are considered to be of the lowest confidence by the SFLD.

All computationally derived annotations rely upon prior knowledge, and so if the rule is not sufficiently detailed, it can still lead to the propagation of annotation errors (*see* Misannotation Section 1.4).

Assigning confidence to annotations is highly subjective [16], however, as one person may consider high-throughput screening, which more frequently is used to predict protein-binding or subcellular locations rather than EC number, of low confidence. This is because such experiments often have a relatively high number of false positives that can generate bias in the analysis. However, depending on what your research questions are, you may consider such data of high confidence. It all depends on what field you are in and what your needs are. Generally speaking, the more reproducible the experiment(s), the higher confidence you can have in their results. Thus, even low-to-medium confident annotations (from Table 1) may lead to a high-confidence annotation.

Table 1
Some example proteins (listed by UniProtKB accession) with their associated annotations, source of the annotation (the SFLD is the Structure-Function Linkage Database, Swiss-Prot is the curated portion of UniProtKB) and the confidence of those annotations along with the reason that confidence level has been assigned

Protein ID from UniProtKB [17]	Annotated protein function (*source*)	SFLD confidence level	Types of evidence or reasoning used to annotate the function
Q9X0Z6	[FeFe]-hydrogenase maturase (*From SFLD and Swiss-Prot*)	High	Inferred from experimental analysis of protein structures, genomic context and results from spectroscopic assay.
Q11S94	Biotin Synthase (BioB) (*From SFLD and Swiss-Prot*)	Medium	Inferred from similarity to other BioB enzymes. Matched by similarity to other BioB sequences and catalytic residues are fully conserved.
Q58692	Biotin Synthase (BioB) (*From Swiss-Prot*)	Low	Inferred from similarity to other BioB enzymes. Matched by similarity to other BioB sequences. Whilst all residues required for binding the iron-sulphur clusters are conserved, all the catalytic residues (those required for the BioB reaction to occur) are not. Also has no biotin synthase genomic context.

For example the GO Reference Code GO_REF:0000003 provides automatic GO annotations based on the mapping of EC numbers to MFO terms, so although annotated as IEA, these annotations can be considered of higher confidence [18]. Some examples of high-, medium- and low-confidence annotations are shown in Table 1, along with reference to the approach used in SwissProt and the SFLD to describe their reliability.

1.3 Annotation Transfer Under the Superfamily Model

We define here an enzyme (or protein) superfamily as the largest grouping of enzymes for which a common ancestry can be identified. Superfamilies can be defined in many different ways, and every resource that utilises them in the bioinformatics community has probably used a slightly different interpretation and method to collate their data. However, they can be broadly classified as structure-based, in which the three-dimensional structures of all available proteins in a superfamily have been aligned and confirmed as homologous, or sequence based, where the sequences have been used rather than structures. Many resources use a combination of approaches. Examples of superfamily based resources include CATH [19], Gene3D [20], SCOP and SUPERFAMILY [21], which are primarily structure based, and Pfam [22], PANTHER [23] and TIGRFAMs [24], which are primarily sequence based. A third definition of a superfamily includes a mechanistic component, i.e. a set of sequences must not only be homologous, but there must be some level of conserved chemical capability within the set, e.g. catalytic residues, cofactors, substrate and/or product substructures or mechanistic steps. An example of such a resource is the SFLD and we will focus on this resource with respect to evaluating GO annotations for enzymes that are members of a defined superfamily.

The SFLD (http://sfld.rbvi.ucsf.edu/) is a manually curated classification resource describing structure-function relationships for functionally diverse enzyme superfamilies [25]. Members of such superfamilies are diverse in their overall reactions yet share a common ancestor and some conserved active site features associated with conserved functional attributes such as a partial reaction or molecular subgraph that all substrates or products may have in common. Thus, despite their different functions, members of these superfamilies often "look alike" which can make them particularly prone to misannotation. To address this complexity and enable reliable transfer of functional features to unknowns only for those members for which we have sufficient functional information, we subdivide superfamily members into subgroups using sequence information (and where available, structural information), and lastly into families, defined as sets of enzymes known to catalyse the same reaction using the same mechanistic strategy and catalytic machinery. At each level of the hierarchy, there are conserved chemical capabilities, which include one or more of the conserved key residues that are responsible for the catalysed function; the small molecule subgraph that all the substrates (or products) may include and any conserved partial reactions. A subgroup is essentially created by observing *a similarity* threshold at which all members of the subgroup have more in common with one another than they do with members of another subgroup. (Thresholds derived from similarity calculations can use many different metrics, such as simple database search programs like BLAST [26] or Hidden Markov Models (HMMs) [27] generated as part of the curation protocol to describe a subgroup or family.)

1.4 Annotation Transfer and Misannotation

Annotation transfer is a hard problem to solve, partly because it is not always easy to know exactly how a function should be transferred. Oftentimes, function and sequence similarity do not track well [28, 29] and so, if sequence similarity is the only criterion that has been used for annotation transfer, the inference of function may have low confidence. However, it is also very difficult to say whether a protein is truly misannotated, especially if no fairly similar protein has been experimentally characterised that could be used for comparison and evaluation of functional features such as the presence of similar functionally important active site residues. As we have previously shown [30–32] there is a truly staggering amount of protein space that has yet to be explored experimentally and that makes it very difficult to make definitive statements as to the validity of an annotation.

Misannotation can come from many sources, from a human making an error in curation, which is then propagated from the top down, to an automated annotation transfer rule that is slightly too lax, to the use of transitivity to transfer annotation, e.g. where protein A is annotated with function X, protein B is 70 % identical to A, and so is also assigned function X, protein C is 65 % identical to protein B, and so is also assigned function X. Whilst this may be the correct function, protein C may have a much lower similarity to protein A, and thus the annotation transfer may be "risky" [33]. As in the example shown in Fig. 3, sequence similarity networks (SSNs) [34] offer a powerful way to highlight where potential

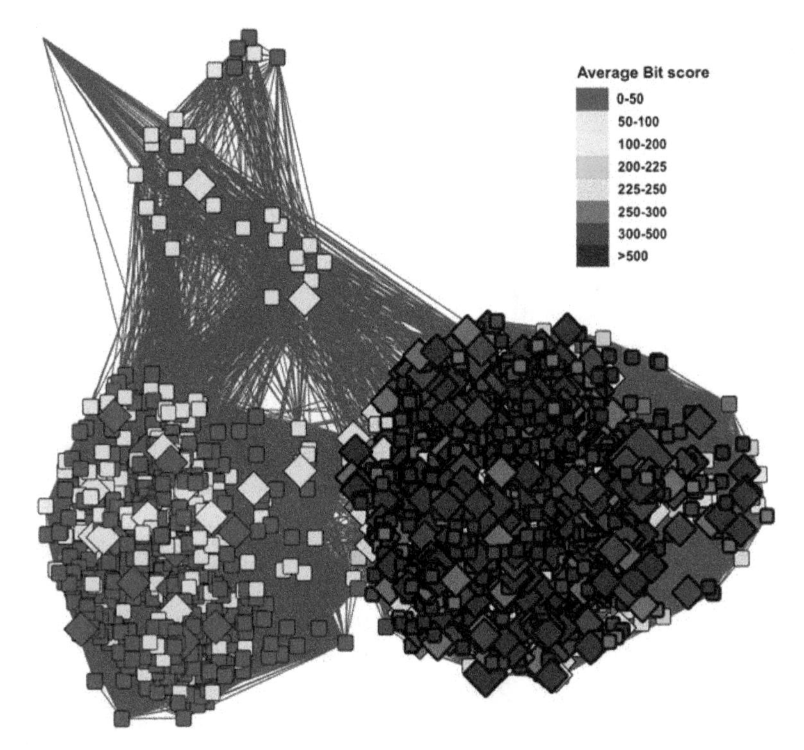

Fig. 3 Example of identifying misannotation using an SSN in the biotin synthase-like subgroup in the SFLD. Nodes colours represent different families in the subgroup, where *red* represent those sets of sequences annotated as canonical biotin synthase in the SFLD, *blue* represent the HydE sequences, *green* the PylB sequences and magenta the HmdB sequences. The nodes shown as *large diamonds* are those annotated as BioB in GO, clearly showing that the annotation transfer for BioB is too broad. The network summarizes the similarity relationships between 5907 sequences. It consists of 2547 representative nodes (nodes represent proteins that share greater than 90 % identity) and 2,133,749 edges, where an edge is the average similarity of pairwise BLAST *E*-values between all possible pairs of the sequences within the connected nodes. In this case, edges are included if this average is more significant than an *E*-value of 1e-25. The organic layout in Cytoscape 3.2.1 is used for graphical depiction. Subheading 2.1 described how such similarity networks are created

misannotation may occur. In this network, all the nodes are connected via a homologous domain, the Radical SAM domain. Thus, the observed differences in the rest of the protein mean that the functions of the proteins may also be quite different. For details on the creation of SSNs, *see* Subheading 2.1. Cases where annotations may be suspect can often be evaluated based on a protein's assigned name, and from the GO terms inferred for that protein.

Not all annotations are created equal, even amongst experimentally validated annotations, and it is important to consider how well evidence supporting an annotation should be trusted. For example, in the glutathione transferase (GST) superfamily, the cognate reaction is often not known as the assays performed use a relatively standard set on non-physiological substrates to infer the type of reaction catalysed by each enzyme that is studied. Moreover, GSTs are often highly promiscuous for two or more different reactions again complicating function assignment [32]. That being said, the availability of even a small amount of experimental evidence can help guide future experiments aimed at functional characterisation. A new ontology, the Confidence Information Ontology (CIO) [16], aims to help annotators assign confidence to evidence. For example, evidence that has been reproduced from many different experiments may have an intrinsically higher confidence than evidence that has only been reported once.

2 Using GO Annotations to Visualise Data in Sequence Similarity Networks

Sequence similarity networks (SSNs) are a key tool that we use in the Structure-Function Linkage Database (SFLD) as they give an immediately accessible view of the superfamily and the relationships between proteins in this set. This in turn allows a user to identify boundaries at which they might reasonably expect to see proteins performing a similar function in a similar manner. As was shown in Fig. 3, the GO annotation for BioB covered several different SFLD families. These annotation terms have been assigned through a variety of methods, but mostly inferred from electronic annotation (i.e. rule-based annotation transfer as shown in Fig. 4).

From the networks shown previously, a user may intuitively see that there are three basic groups of proteins. Further, it could be hypothesised that these groups could have different functions (which is indeed the case in this particular example). Thus, the user may be left with the question: How do I know what boundaries to use for high confidence in the annotation transfer? Figure 5 shows another network, this time coloured by the average bit-score for the sequences in a node against the SFLD HMM for BioB. This network exemplifies how (1) sequence similarity (network clusters) corresponds with the sequence pattern generated by SFLD curators to represent the BioB family, and (2) HMM true-positive gathering

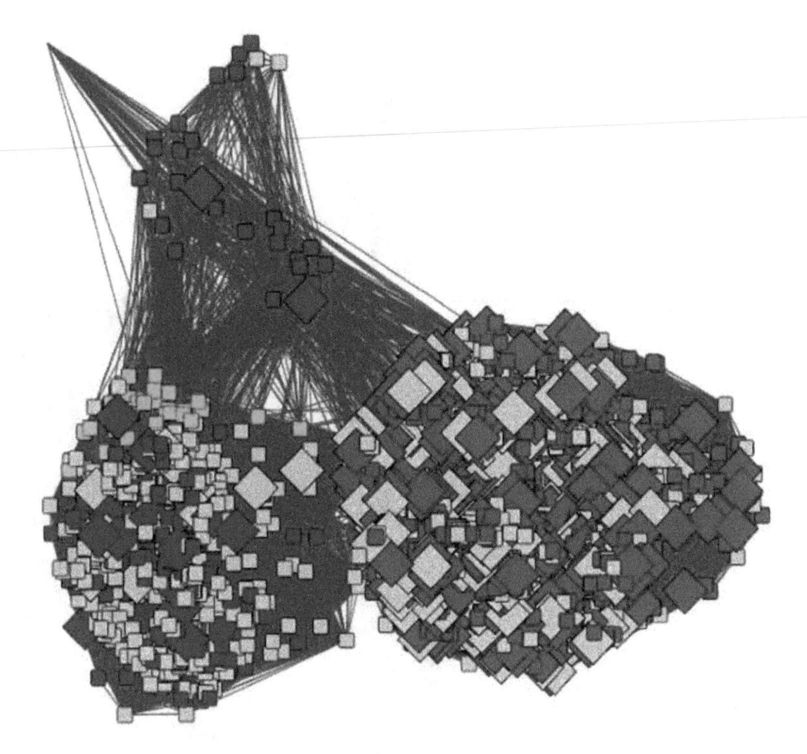

Fig. 4 Biotin synthase-like subgroup coloured by confidence of evidence (as shown in Table 1). The *diamond shaped nodes* are all annotated as Biotin Synthase in GO. *Red nodes* are those that only have low confidence annotations, the *orange nodes* are those that have at least one medium-confidence annotations and the *green* are those that have at least one high-confidence annotation. *Grey nodes* have no BioB annotations. Node and edge numbers, as well as *e*-value threshold are as in Fig. 3

bit-score cut-off can be fine-tuned. By combining what we know about the protein set from the GO annotation (Fig. 3) with the HMM bit-score (Fig. 5) it is possible to be much more confident in the annotations for the proteins in the red/brown group in Fig. 5.

2.1 Creating Sequence Similarity Networks

SSNs provide a visually intuitive method for viewing large sets of similarities between proteins [34]. Although their generation is subject to size limitations for truly large data sets, they can be easily created and visualised for several thousand sequences. There are many ways to create such networks, the networks created by the SFLD are generated by Pythoscape [35], a freely available software that can be downloaded, installed and can be run locally. Recently, web servers have been described that will generate networks for users. For example, The Enzyme Similarity Tool (EFI-EST) [36] created by the Enzyme Function Initiative will take a known set of proteins (e.g. Pfam or InterPro [37] groups) and generate networks for users from that set. A similarity network is simply a set of nodes (representing a set of amino acid sequences as described in

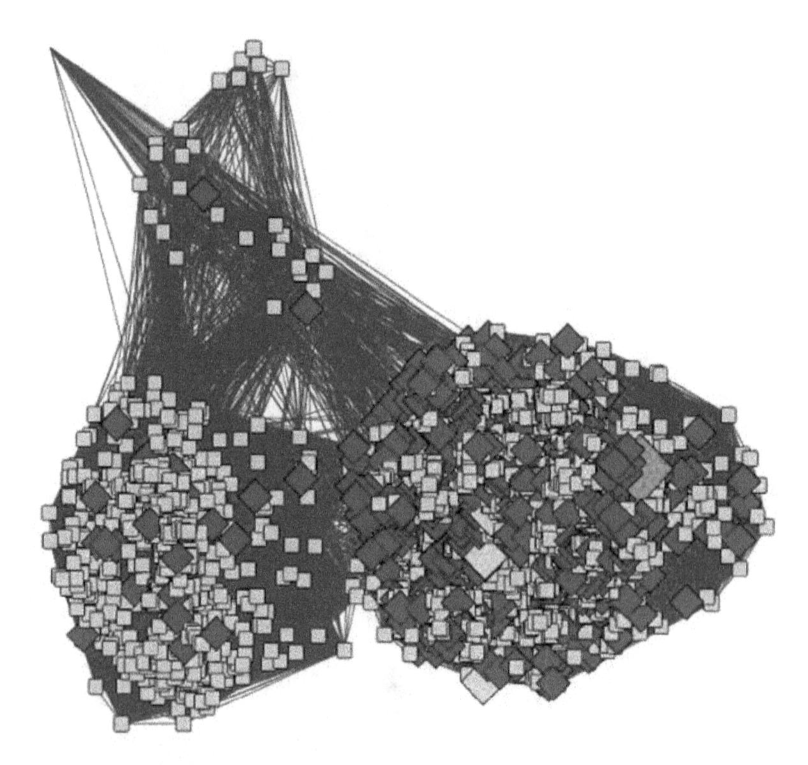

Fig. 5 Example of a sequence similarity network to estimate subgroups for use in initial steps of the curation process and to guide fine-tuning the hidden Markov model (HMM) true-positive detection threshold of an enzyme family (here for the Biotin synthase (BioB) family). Node colours represent the average Bit-score of the BioB family HMM for all sequences represented by the node. The mapping between colours and average Bit scores is given in the legend. Nodes with *thick borders* represent proteins that belong to the BioB family according to SFLD annotation. *Diamonds* represent nodes that include proteins with BioB family annotation according to GO. The final BioB HMM detection threshold was achieved for the SFLD by further exploration of more strict *E*-value thresholds for edge representation, and was set to 241.6. Node and edge numbers, as well as *E*-value threshold, are as in Fig. 3

this chapter, for example) and edges (representing the similarity between those nodes). For the SSNs shown in this chapter, edges represent similarities scored by pairwise BLAST *E*-values (used as scores) between the source and target sequences. Using simple metrics such as these, relatively small networks are trivial and fast to produce from a simple all-against-all BLAST calculation. However, the number of edges produced depends on the similarity between all the nodes to each other, so that for comparisons of a large number of closely related sequences, the number of edges will vastly exceed the number of nodes, quickly outpacing computational resources for generating and viewing networks. As a result, some data reduction will eventually be necessary. The SFLD uses representative networks where each node represents a set of highly similar sequences and the edges between them represent the mean *E*-value similarity between all the sequences in the source node and all the sequences in the target node. As shown in Fig. 3, node graphical attributes (e.g. shape and colour) used to represent GO terms for the proteins shown are a powerful way to recognise relationships between sequence and functional similarities. Importantly, statistical analyses must be carried out to verify the significance of these trends, as we show below.

2.2 Determining Over- and Under-represented GO Terms in a Set of Species-Diverse Proteins

A common use of GO enrichment analysis is to evaluate sets of differentially expressed genes that are up- or down-regulated under certain conditions [38]. The resulting analysis identifies which GO terms are over- or under-represented within the set in question. With respect to enzyme superfamilies, the traditional implementation of enrichment analysis will not work well as there are often very many different species from different kingdoms in the dataset. However, there are several ways that we can still utilise sets of annotated proteins to evaluate the level of enrichment for GO terms.

The simplest method and least rigorous, is to take the set of proteins being evaluated, count up the number of times a single annotation occurs (including duplicate occurrences for a single enzyme, as these have different evidence sources) and up-weight for experimental (or high confidence) annotations. Then, by dividing by the number of proteins in the set, any annotation with a ratio greater than one can be considered “significant”.

A more rigorous treatment assumes that for a set of closely related proteins (i.e. belonging to a family) a specific GO term is said to be over-represented when the number of proteins assigned to that term within the family of interest is enriched versus the background model as determined by a probability distribution. Thus, there are two decisions that need to be made, firstly, identifying the background model and then which probability function to use. The background model is dependent on the dataset and the question that is being asked. For example in the SFLD model, we might use the subgroup or superfamily and a random background model that gives us an idea of what annotations could occur purely by chance. The lack of high (and sometimes also medium) confidence annotations is another complication in examining enrichment of terms. If one is using IEA annotations to infer function, the assertions can quickly become circular (with inferred annotations being transferred to other proteins which in turn are used to annotate yet more proteins), leading to results which themselves are of low confidence. Similarly, if very few proteins are explicitly annotated with a high/medium confidence annotation, the measure of significance can be skewed due to low counts in the dataset. The choice of the probability function is also going to depend somewhat on what question is being asked, but the hypergeometric test (used for a finite universe) is common in GO analyses [39, 40]. For more detail on enrichment analysis, *see* Chap. 13 [41].

2.3 Using Semantic Significance with GO

Instead of simply transferring annotations utilising sequence homology and BLAST scores, many tools are now available (e.g. Argot2 [42] and GraSM [43]) that utilise semantic similarity [42–46]. Here, the idea is that in controlled vocabularies, the degree of relatedness between two entities can be assessed by comparing the semantic relationship (meanings) between their annotations. The semantic similarity measure is returned as a numerical value that quantifies the relationship between two GO terms, or two sets of terms annotating two proteins.

GO is well suited to such an approach, for example many children terms in the GO directed acyclic graph (DAG) have a similar vocabulary to their parents. The nature of the GO DAG means that a protein with a function A will also inherit the more generic functions that appear higher up in the DAG; this can be one or more functions, depending on the DAG. For example, an ion transmembrane transporter activity (GO:0015075) is a term similar to voltage-gated ion channel activity (GO:0005244), the latter of which is a descendent of the former, albeit separated by the ion channel activity (GO:0005216) term. Thus, the ancestry and semantic similarity lends greater weight to the confidence in the annotation.

Such similarity measures can be used instead of (or in conjunction with) sequence similarity measures. Indeed, it has been shown [47] that there is good correlation between the protein sequence similarity and the GO annotation semantic similarity for proteins in Swiss-Prot, the reviewed section of UniProtKB [17]. Consistent results, however, are often a feature not only of the branch of GO to which the annotations belong, but also the number of high confidence annotations that are being used. For a more detailed and comprehensive discussion of the various methods, *see* Pesquita et al. [44] and Chap. 12 [48].

2.4 Use of Orthogonal Information to Evaluate GO Annotation

In the example shown in Fig. 3, it is clear that many more nodes in the subgroup are annotated as biotin synthase by GO than match the stringent criteria set within the SFLD, which not only require a significant *E*-value (or Bit Score) to transfer annotation, but the presence of the conserved key residues. As mentioned earlier, one key advantage to using GO annotations over those of some other resources is the evidence code (and associated source of that evidence) as shown in Fig. 4. As indicated by that network, when using GO annotations, it is important to also consider the associated confidence level for the evidence used in assigning an annotation (*see* Table 1). In Fig. 4, only a few annotations are supported by high-confidence evidence. Alternatively, if a protein has a high confidence experimental evidence code for membership in a family of interest yet is not included by annotators in that family, then the definition of that family may be too strict, indicating that a more permissive gathering threshold for assignment to the family should be used.

Another way of assessing the veracity of the annotation transferred to a query protein is to examine both the annotations of the proteins that are closest to it in similarity as well as other entirely different types of information.

One example of such orthogonal information is the genomic context of the protein. It can be hypothesised that if a protein occurs in a pathway, then the other proteins involved in that pathway may be co-located within the genome [49]. This association is frequently found in prokaryotes, and to a lesser extent in plants and fungi. Genomic proximity of pathway components is infrequent in metazoans, thus genomic context as a means to function prediction is more useful for bacterial enzymes. Additionally, other genes in the same genomic neighbourhood may be relevant to understanding the function of both the protein of interest and of the associated pathway. A common genomic context for a query protein and a homologue provides further support for assignment of that function. (However, the genomic distance between pathway components in different organisms may vary for many reasons, thus the lack of similar genomic context does not suggest that the functions of a query and a similar homologue are different.)

Another type of orthogonal information that can be used can be deduced from protein domains present in a query protein and their associated annotations—what are the predicted domains present in the protein, do they all match the assigned function or are there anomalies. A good service for identifying such domains is InterProScan [50]. Further, any protein in UniProtKB will have the predicted InterPro identifiers annotated in the record (along with other predicted annotations from resources such as Pfam and CATH), along with the evidence supporting those predictions. Such sequence context can also be obtained using hidden Markov models (HMMs) [51], which is the technique used by InterPro, Pfam, Gene3D, SUPERFAMILY and the SFLD to place new sequences into families, subgroups (SFLD-specific term) and superfamilies (*see* Fig. 6).

3 Challenges and Caveats

3.1 The Use of Sequence Similarity Network

A significant challenge with using SSNs to help evaluate GO annotations is that SSNs are not always trivial to use without a detailed knowledge of the superfamilies that they describe. For example, choosing an appropriate threshold for drawing edges is critical to obtaining network clustering patterns useful for deeper evaluation. In Fig. 3, HydE (the blue nodes) are not currently annotated as such in GO, but are annotated instead as BioB. Thus,

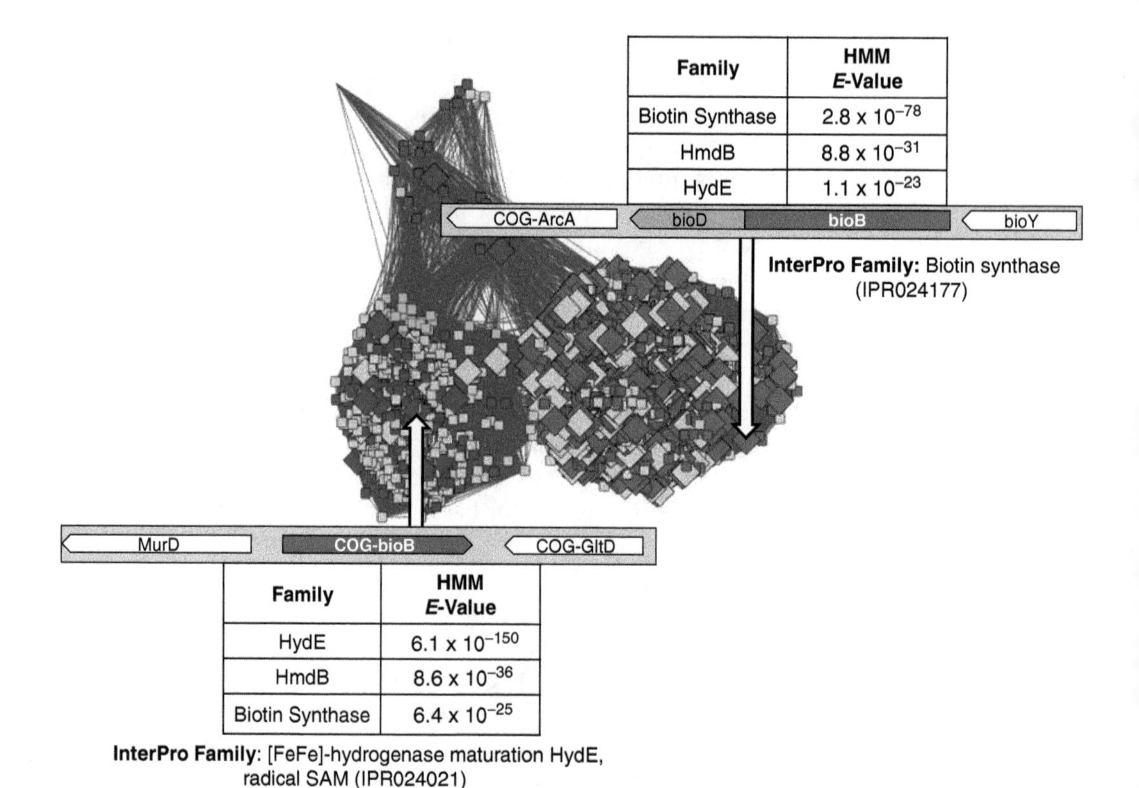

Fig. 6 Biotin synthase-like subgroup SSN showing where the biotin synthase GO annotations are shown as *large diamonds*. Two proteins, one from the BioB set (*red nodes, top right*) and one from the HydE set (*blue nodes*), *bottom left*, are shown with some associate orthogonal information: genomic context highlighted in *light cyan boxes*, their HMM match results for the query protein against the three top scoring families in the subgroup are shown in the tables, and family membership (according to InterProScan) shown in coloured text (*blue* for HydE and *red* for BioB). Node and edge numbers, as well as *E*-value threshold are as in Fig. 3. All the proteins are connected via a homologous domain (the Radical SAM domain). Thus, the observed differences in the rest of the protein mean that the functions of the proteins may also be quite different

the evaluation of the network becomes significantly more complex. It is also not always clear what signal is being picked up in the edge data for large networks. It is usually assumed that all the proteins in the set share a single domain, but this is often only clear when the network is examined in greater detail.

3.2 Annotation Transfer Is Challenging Because Evolution Is Complex

Even using the powerful tools and classifications provided by GO, interpreting protein function in many cases requires more in-depth analysis. For several reasons, it is not always easy to confidently determine that a protein is not correctly annotated. Firstly, how closely related is the enzyme to the group of interest? Perhaps we can only be relatively certain of its superfamily membership, or maybe we can assign it to a more detailed level of the functional hierarchy. If it fits into a more detailed classification level, how well does it fit? At what threshold do we begin to see false positives creeping into the results list? Using networks, we can also examine the closest neighbours that have differing function and ask whether there are similarities in the function (e.g. Broderick et al. [52] used sequence similarity networks to help determine the function of HydE). Another complicating issue is whether a protein performs one or more promiscuous functions, albeit with a lesser efficacy.

Another important piece of evidence that can be used to support an annotation is conservation of the key residues, so it is important to assess if the protein of interest has all the relevant functional residues. Although GO includes an evidence code to handle this concept (Inferred from Key Residues, IKR), it is often not included in the electronic inference of annotations. It is important to note, however, that there are evolutionary events that may "scramble" the sequence, leaving it unclear to an initial examination whether the residues are conserved or not. A prime example is the case in which a circular permutation has occurred. Thus, it is important to look at whether there are other residues (or patterns of residues) that could perform the function of the "missing" residues. It is also possible that conservative mutations have occurred, and these may also have the ability to perform the function of the "missing" residues [53].

Another consideration with function evaluation is the occurrence of moonlighting proteins. These are proteins that are identical in terms of sequence but perform different functions in different cellular locations or species; for example argininosuccinate lyase (UniProtKB id P24058) is also a delta crystalline which serves as an eye lens protein when it is found in birds and reptiles [54]. A good source of information on moonlighting proteins is MoonProt (http://www.moonlightingproteins.org/) [55]. Such cases may arise from physiological use in many different conditions such as different subcellular localisations or regulatory pathways. The full extent of proteins that moonlight is currently not known, although to date, almost 300 cases have been reported in MoonProt. Another complicating factor for understanding the evolution of enzyme function is the apparent evolution of the same reaction specificity from different intermediate nodes in the phylogenetic tree for the superfamily, for example the N-succinyl amino acid racemase and the muconate lactonising enzyme families in the enolase superfamily [56, 57].

Finally, does the protein have a multi-domain architecture and/or is it part of a non-covalent protein-protein interaction in the cell? An example of a functional protein requiring multiple chains that are transiently coordinated in the cell is pyruvate dehydrogenase (acetyl-transferring) (EC 1.2.4.1). This protein has an active site at the interface between pyruvate dehydrogenase E1 component subunit alpha (UniProtKB identifier P21873) and beta (UniProtKB identifier P21874), both of which are required for activity. Thus, transfer of annotation relating to this function to an unknown (and hence evaluation of misannotation) needs to include both proteins. Similarly, a single chain with multiple domains, e.g. biotin biosynthesis bifunctional protein BioAB (UniProtKB identifier P53656), which contains a BioA and BioB domain, has two different functions associated with it. In this example, these two functions are distinct from one another so that annotation of this protein only with one function or the other could represent a type of misannotation (especially as a GO term is assigned to a protein, not a specific segment of its amino acid sequence).

3.3 Plurality Vote May Not Be the Best Route

In some cases, proteins are annotated by some type of "plurality voting". Plurality voting is simply assuming that the more annotations that come from different predictors, the more likely these are to be correct. As we have shown in this chapter (and others before us [58]), this is not always the case. An especially good example of where plurality voting fails is in the case of the lysozyme mechanism. For over 50 years, the mechanism was assumed to be dissociative, but a single experiment provided evidence of a covalent intermediate being formed in the crystal structure, calling into question the dissociate mechanism. If plurality voting were applied in ongoing annotations, the old mechanism would still be considered correct. That being said, it is more difficult to identify problems of this type if experimental evidence challenging an annotation is unavailable. In such cases, we must always look at all the available evidence to transfer function and where there are disagreements between predicted functions, a more detailed examination is needed. Only when we have resolved such issues can we have any true confidence in the plurality vote. Work by Kristensen et al. [59] provides a good example of the value of this approach. By using three-dimensional templates generated using knowledge of the evolutionarily important residues, they showed that they could identify a single most likely function in 61% of 3D structures from the Structural Genomics Initiative, and in those cases the correct function was identified with an 87% accuracy.

4 Conclusions

Experimentalists simply can't keep up with the huge volume of data that is being produced in today's high-throughput labs, from whole genome and population sequencing efforts to large-scale assays and structure generation. Almost all proteins will have at least one associated GO annotation, and such coverage makes GO an incredibly powerful tool, especially as it has the ability to handle all the known function information at different levels of biological granularity, has explicit tools to capture high-throughput experimental data and utilises an ontology to store the annotation and associated relationships. Although over 98% of all GO annotations are computationally inferred, with the ever-increasing state of knowledge, these annotation transfers are becoming more confident [15] as rule-based annotations gain in specificity due to more data being available. However, there is still a long way to go before we can simply take an IEA annotation at face value. Confidence in annotations transferred electronically has to be taken into account: How many different sources have come to the same conclusion (using different methods)? How many different proteins' functions have been determined in a single experiment? Similarly, whilst burden of evidence is a useful gauge in determining the significance of an annotation, there is also the question of when substantially different annotations were captured in GO and other resources—perhaps there has been a new experiment that calls into question the original annotation. It is also important to look at whether other, similar proteins were annotated long ago or are based on new experimental evidence. There is a wealth of data available that relates to enzymes and their functions. This ranges from the highest level of associating a protein with a superfamily (and thus giving some information as to the amino acid residues that are evolutionarily conserved), to the most detailed level of molecular function. We can use all of these data to aid us in evaluating the GO annotations for a given protein (or set of proteins), from the electronically inferred annotation for protein domain structure, to the genomic context and protein features (such as conserved residues). The more data that are available to back up (or refute) a given GO annotation, the more confident one can be in it (or not, as the case may be).

Acknowledgements

GLH and PCB acknowledge funds from National Institutes of Health and National Science Foundation (grant NIH R01 GM60595 and grant NSF DBI1356193). Open Access charges were funded by the University College London Library, the Swiss Institute of Bioinformatics, the Agassiz Foundation, and the Foundation for the University of Lausanne.

References

1. Ashburner M, Ball CA, Blake JA, Botstein D, Butler H, Cherry JM, Davis AP, Dolinski K, Dwight SS, Eppig JT, Harris MA, Hill DP, Issel-Tarver L, Kasarskis A, Lewis S, Matese JC, Richardson JE, Ringwald M, Rubin GM, Sherlock G (2000) Gene ontology: tool for the unification of biology. The Gene Ontology Consortium. Nat Genet 25(1):25–29. doi:10.1038/75556
2. Nomenclature committee of the international union of biochemistry and molecular biology (NC-IUBMB), Enzyme Supplement 5 (1999). European J Biochem/FEBS 264(2):610–650
3. McDonald AG, Boyce S, Tipton KF (2009) ExplorEnz: the primary source of the IUBMB enzyme list. Nucleic Acids Res 37(Database issue):D593–D597. doi:10.1093/nar/gkn582
4. Fleischmann A, Darsow M, Degtyarenko K, Fleischmann W, Boyce S, Axelsen KB, Bairoch A, Schomburg D, Tipton KF, Apweiler R (2004) IntEnz, the integrated relational enzyme database. Nucleic Acids Res 32(Database issue):D434–D437. doi:10.1093/nar/gkh119

5. Furnham N (2016) Complementary sources of protein functional information: the far side of GO. In: Dessimoz C, Škunca N (eds) The gene ontology handbook. Methods in molecular biology, vol 1446. Humana Press. Chapter 19
6. Babbitt PC (2003) Definitions of enzyme function for the structural genomics era. Curr Opin Chem Biol 7(2):230–237
7. Thomas PD (2016) The gene ontology and the meaning of biological function. In: Dessimoz C, Škunca N (eds) The gene ontology handbook. Methods in molecular biology, vol 1446. Humana Press. Chapter 2
8. Bray T, Doig AJ, Warwicker J (2009) Sequence and structural features of enzymes and their active sites by EC class. J Mol Biol 386(5):1423–1436. doi:10.1016/j.jmb.2008.11.057
9. Dobson PD, Doig AJ (2005) Predicting enzyme class from protein structure without alignments. J Mol Biol 345(1):187–199. doi:10.1016/j.jmb.2004.10.024
10. Cozzetto D, Jones DT (2016) Computational methods for annotation transfers from sequence. In: Dessimoz C, Škunca N (eds) The gene ontology handbook. Methods in molecular biology, vol 1446. Humana Press. Chapter 5
11. Radivojac P, Clark WT, Oron TR, Schnoes AM, Wittkop T, Sokolov A, Graim K, Funk C, Verspoor K, Ben-Hur A, Pandey G, Yunes JM, Talwalkar AS, Repo S, Souza ML, Piovesan D, Casadio R, Wang Z, Cheng J, Fang H, Gough J, Koskinen P, Toronen P, Nokso-Koivisto J, Holm L, Cozzetto D, Buchan DW, Bryson K, Jones DT, Limaye B, Inamdar H, Datta A, Manjari SK, Joshi R, Chitale M, Kihara D, Lisewski AM, Erdin S, Venner E, Lichtarge O, Rentzsch R, Yang H, Romero AE, Bhat P, Paccanaro A, Hamp T, Kassner R, Seemayer S, Vicedo E, Schaefer C, Achten D, Auer F, Boehm A, Braun T, Hecht M, Heron M, Honigschmid P, Hopf TA, Kaufmann S, Kiening M, Krompass D, Landerer C, Mahlich Y, Roos M, Bjorne J, Salakoski T, Wong A, Shatkay H, Gatzmann F, Sommer I, Wass MN, Sternberg MJ, Skunca N, Supek F, Bosnjak M, Panov P, Dzeroski S, Smuc T, Kourmpetis YA, van Dijk AD, ter Braak CJ, Zhou Y, Gong Q, Dong X, Tian W, Falda M, Fontana P, Lavezzo E, Di Camillo B, Toppo S, Lan L, Djuric N, Guo Y, Vucetic S, Bairoch A, Linial M, Babbitt PC, Brenner SE, Orengo C, Rost B, Mooney SD, Friedberg I (2013) A large-scale evaluation of computational protein function prediction. Nat Methods 10(3):221–227. doi:10.1038/nmeth.2340
12. Friedberg I, Radivojac P (2016) Community-wide evaluation of computational function prediction. In: Dessimoz C, Škunca N (eds) The gene ontology handbook. Methods in molecular biology, vol 1446. Humana Press. Chapter 10
13. Akiva E, Brown S, Almonacid DE, Barber AE 2nd, Custer AF, Hicks MA, Huang CC, Lauck F, Mashiyama ST, Meng EC, Mischel D, Morris JH, Ojha S, Schnoes AM, Stryke D, Yunes JM, Ferrin TE, Holliday GL, Babbitt PC (2014) The Structure-Function Linkage Database. Nucleic Acids Res 42(Database issue):D521–D530. doi:10.1093/nar/gkt1130
14. Gaudet P, Škunca N, Hu JC, Dessimoz C (2016) Primer on the gene ontology. In: Dessimoz C, Škunca N (eds) The gene ontology handbook. Methods in molecular biology, vol 1446. Humana Press. Chapter 3
15. Skunca N, Altenhoff A, Dessimoz C (2012) Quality of computationally inferred gene ontology annotations. PLoS Comput Biol 8(5):e1002533. doi:10.1371/journal.pcbi.1002533
16. Bastian FB, Chibucos MC, Gaudet P, Giglio M, Holliday GL, Huang H, Lewis SE, Niknejad A, Orchard S, Poux S, Skunca N, Robinson-Rechavi M (2015) The Confidence Information Ontology: a step towards a standard for asserting confidence in annotations. Database:bav043. doi:10.1093/database/bav043
17. UniProt C (2015) UniProt: a hub for protein information. Nucleic Acids Res 43(Database issue):D204–D212. doi:10.1093/nar/gku989
18. Hill DP, Davis AP, Richardson JE, Corradi JP, Ringwald M, Eppig JT, Blake JA (2001) Program description: strategies for biological annotation of mammalian systems: implementing gene ontologies in mouse genome informatics. Genomics 74(1):121–128. doi:10.1006/geno.2001.6513
19. Sillitoe I, Lewis TE, Cuff A, Das S, Ashford P, Dawson NL, Furnham N, Laskowski RA, Lee D, Lees JG, Lehtinen S, Studer RA, Thornton J, Orengo CA (2015) CATH: comprehensive structural and functional annotations for genome sequences. Nucleic Acids Res 43(Database issue):D376–D381. doi:10.1093/nar/gku947
20. Lees J, Yeats C, Perkins J, Sillitoe I, Rentzsch R, Dessailly BH, Orengo C (2012) Gene3D: a domain-based resource for comparative genomics, functional annotation and protein network analysis. Nucleic Acids Res 40(Database issue):D465–D471. doi:10.1093/nar/gkr1181

21. Fox NK, Brenner SE, Chandonia JM (2014) SCOPe: structural classification of proteins--extended, integrating SCOP and ASTRAL data and classification of new structures. Nucleic Acids Res 42(Database issue):D304–D309. doi:10.1093/nar/gkt1240
22. Finn RD, Bateman A, Clements J, Coggill P, Eberhardt RY, Eddy SR, Heger A, Hetherington K, Holm L, Mistry J, Sonnhammer EL, Tate J, Punta M (2014) Pfam: the protein families database. Nucleic Acids Res 42(Database issue):D222–D230. doi:10.1093/nar/gkt1223
23. Mi H, Muruganujan A, Thomas PD (2013) PANTHER in 2013: modeling the evolution of gene function, and other gene attributes, in the context of phylogenetic trees. Nucleic Acids Res 41(Database issue):D377–D386. doi:10.1093/nar/gks1118
24. Haft DH, Selengut JD, Richter RA, Harkins D, Basu MK, Beck E (2013) TIGRFAMs and genome properties in 2013. Nucleic Acids Res 41(Database issue):D387–D395. doi:10.1093/nar/gks1234
25. Gerlt JA, Babbitt PC (2001) Divergent evolution of enzymatic function: mechanistically diverse superfamilies and functionally distinct suprafamilies. Annu Rev Biochem 70:209–246. doi:10.1146/annurev.biochem.70.1.209
26. Camacho C, Coulouris G, Avagyan V, Ma N, Papadopoulos J, Bealer K, Madden TL (2009) BLAST+: architecture and applications. BMC Bioinformatics 10:421. doi:10.1186/1471-2105-10-421
27. Finn RD, Clements J, Eddy SR (2011) HMMER web server: interactive sequence similarity searching. Nucleic Acids Res 39(Web Server issue):W29–W37. doi:10.1093/nar/gkr367
28. Brown SD, Babbitt PC (2014) New insights about enzyme evolution from large scale studies of sequence and structure relationships. J Biol Chem 289(44):30221–30228. doi:10.1074/jbc.R114.569350
29. Schnoes AM, Brown SD, Dodevski I, Babbitt PC (2009) Annotation error in public databases: misannotation of molecular function in enzyme superfamilies. PLoS Comput Biol 5(12):e1000605. doi:10.1371/journal.pcbi.1000605
30. Pieper U, Chiang R, Seffernick JJ, Brown SD, Glasner ME, Kelly L, Eswar N, Sauder JM, Bonanno JB, Swaminathan S, Burley SK, Zheng X, Chance MR, Almo SC, Gerlt JA, Raushel FM, Jacobson MP, Babbitt PC, Sali A (2009) Target selection and annotation for the structural genomics of the amidohydrolase and enolase superfamilies. J Struct Funct Genom 10(2):107–125. doi:10.1007/s10969-008-9056-5
31. Gerlt JA, Babbitt PC, Jacobson MP, Almo SC (2012) Divergent evolution in enolase superfamily: strategies for assigning functions. J Biol Chem 287(1):29–34. doi:10.1074/jbc.R111.240945
32. Mashiyama ST, Malabanan MM, Akiva E, Bhosle R, Branch MC, Hillerich B, Jagessar K, Kim J, Patskovsky Y, Seidel RD, Stead M, Toro R, Vetting MW, Almo SC, Armstrong RN, Babbitt PC (2014) Large-scale determination of sequence, structure, and function relationships in cytosolic glutathione transferases across the biosphere. PLoS Biol 12(4):e1001843. doi:10.1371/journal.pbio.1001843
33. Rentzsch R, Orengo CA (2013) Protein function prediction using domain families. BMC Bioinformatics 14(Suppl 3):S5. doi:10.1186/1471-2105-14-S3-S5
34. Atkinson HJ, Morris JH, Ferrin TE, Babbitt PC (2009) Using sequence similarity networks for visualization of relationships across diverse protein superfamilies. PLoS One 4(2):e4345. doi:10.1371/journal.pone.0004345
35. Barber AE II, Babbitt PC (2012) Pythoscape: a framework for generation of large protein similarity networks. Bioinformatics. doi:10.1093/bioinformatics/bts532
36. Gerlt JA, Bouvier JT, Davidson DB, Imker HJ, Sadkhin B, Slater DR, Whalen KL (2015) Enzyme Function Initiative-Enzyme Similarity Tool (EFI-EST): a web tool for generating protein sequence similarity networks. Biochim Biophys Acta 1854(8):1019–1037. doi:10.1016/j.bbapap.2015.04.015
37. Mitchell A, Chang HY, Daugherty L, Fraser M, Hunter S, Lopez R, McAnulla C, McMenamin C, Nuka G, Pesseat S, Sangrador-Vegas A, Scheremetjew M, Rato C, Yong SY, Bateman A, Punta M, Attwood TK, Sigrist CJ, Redaschi N, Rivoire C, Xenarios I, Kahn D, Guyot D, Bork P, Letunic I, Gough J, Oates M, Haft D, Huang H, Natale DA, Wu CH, Orengo C, Sillitoe I, Mi H, Thomas PD, Finn RD (2014) The InterPro protein families database: the classification resource after 15 years. Nucleic Acids Res. doi:10.1093/nar/gku1243
38. Webber C (2011) Functional enrichment analysis with structural variants: pitfalls and strategies. Cytogenet Genome Res 135(3-4):277–285. doi:10.1159/000331670
39. Thomas PD, Wood V, Mungall CJ, Lewis SE, Blake JA, Gene Ontology C (2012) On the use of gene ontology annotations to assess functional similarity among orthologs and paralogs: a short report. PLoS Comput Biol 8(2):e1002386. doi:10.1371/journal.pcbi.1002386

40. Cao J, Zhang S (2014) A Bayesian extension of the hypergeometric test for functional enrichment analysis. Biometrics 70(1):84–94. doi:10.1111/biom.12122
41. Bauer S (2016) Gene-category analysis. In: Dessimoz C, Škunca N (eds) The gene ontology handbook. Methods in molecular biology, vol 1446. Humana Press. Chapter 13
42. Falda M, Toppo S, Pescarolo A, Lavezzo E, Di Camillo B, Facchinetti A, Cilia E, Velasco R, Fontana P (2012) Argot2: a large scale function prediction tool relying on semantic similarity of weighted Gene Ontology terms. BMC Bioinformatics 13(Suppl 4):S14. doi:10.1186/1471-2105-13-S4-S14
43. Couto FM, Silva MJ, Coutinho PM (2007) Measuring semantic similarity between Gene Ontology terms. Data Knowl Eng 61(1):137–152. doi:10.1016/j.datak.2006.05.003
44. Pesquita C, Faria D, Falcao AO, Lord P, Couto FM (2009) Semantic similarity in biomedical ontologies. PLoS Comput Biol 5(7):e1000443. doi:10.1371/journal.pcbi.1000443
45. Benabderrahmane S, Smail-Tabbone M, Poch O, Napoli A, Devignes MD (2010) IntelliGO: a new vector-based semantic similarity measure including annotation origin. BMC Bioinformatics 11:588. doi:10.1186/1471-2105-11-588
46. Wu X, Pang E, Lin K, Pei ZM (2013) Improving the measurement of semantic similarity between gene ontology terms and gene products: insights from an edge- and IC-based hybrid method. PLoS One 8(5):e66745. doi:10.1371/journal.pone.0066745
47. Apweiler R, Bairoch A, Wu CH, Barker WC, Boeckmann B, Ferro S, Gasteiger E, Huang H, Lopez R, Magrane M, Martin MJ, Natale DA, O'Donovan C, Redaschi N, Yeh LS (2004) UniProt: the Universal Protein knowledgebase. Nucleic Acids Res 32(Database issue):D115–D119. doi:10.1093/nar/gkh131
48. Pesquita C (2016) Semantic similarity in the gene ontology. In: Dessimoz C, Škunca N (eds) The gene ontology handbook. Methods in molecular biology, vol 1446. Humana Press. Chapter 12
49. Huynen M, Snel B, Lathe W, Bork P (2000) Exploitation of gene context. Curr Opin Struct Biol 10(3):366–370
50. Li W, Cowley A, Uludag M, Gur T, McWilliam H, Squizzato S, Park YM, Buso N, Lopez R (2015) The EMBL-EBI bioinformatics web and programmatic tools framework. Nucleic Acids Res. doi:10.1093/nar/gkv279
51. Meng X, Ji Y (2013) Modern computational techniques for the HMMER sequence analysis. ISRN Bioinformatics 2013:252183. doi:10.1155/2013/252183
52. Betz JN, Boswell NW, Fugate CJ, Holliday GL, Akiva E, Scott AG, Babbitt PC, Peters JW, Shepard EM, Broderick JB (2015) [FeFe]-hydrogenase maturation: insights into the role HydE plays in dithiomethylamine biosynthesis. Biochemistry 54(9):1807–1818. doi:10.1021/bi501205e
53. Wellner A, Raitses Gurevich M, Tawfik DS (2013) Mechanisms of protein sequence divergence and incompatibility. PLoS Genet 9(7):e1003665. doi:10.1371/journal.pgen.1003665
54. Sampaleanu LM, Yu B, Howell PL (2002) Mutational analysis of duck delta 2 crystallin and the structure of an inactive mutant with bound substrate provide insight into the enzymatic mechanism of argininosuccinate lyase. J Biol Chem 277(6):4166–4175. doi:10.1074/jbc.M107465200
55. Mani M, Chen C, Amblee V, Liu H, Mathur T, Zwicke G, Zabad S, Patel B, Thakkar J, Jeffery CJ (2015) MoonProt: a database for proteins that are known to moonlight. Nucleic Acids Res 43(Database issue):D277–D282. doi:10.1093/nar/gku954
56. Song L, Kalyanaraman C, Fedorov AA, Fedorov EV, Glasner ME, Brown S, Imker HJ, Babbitt PC, Almo SC, Jacobson MP, Gerlt JA (2007) Prediction and assignment of function for a divergent N-succinyl amino acid racemase. Nat Chem Biol 3(8):486–491. doi:10.1038/nchembio.2007.11
57. Sakai A, Fedorov AA, Fedorov EV, Schnoes AM, Glasner ME, Brown S, Rutter ME, Bain K, Chang S, Gheyi T, Sauder JM, Burley SK, Babbitt PC, Almo SC, Gerlt JA (2009) Evolution of enzymatic activities in the enolase superfamily: stereochemically distinct mechanisms in two families of cis, cis-muconate lactonizing enzymes. Biochemistry 48(7):1445–1453. doi:10.1021/bi802277h
58. Brenner SE (1999) Errors in genome annotation. Trends Genet 15(4):132–133
59. Kristensen DM, Ward RM, Lisewski AM, Erdin S, Chen BY, Fofanov VY, Kimmel M, Kavraki LE, Lichtarge O (2008) Prediction of enzyme function based on 3D templates of evolutionarily important amino acids. BMC Bioinformatics 9:17. doi:10.1186/1471-2105-9-17

9

Community-Wide Evaluation of Computational Function Prediction

Iddo Friedberg and Predrag Radivojac

Abstract

A biological experiment is the most reliable way of assigning function to a protein. However, in the era of high-throughput sequencing, scientists are unable to carry out experiments to determine the function of every single gene product. Therefore, to gain insights into the activity of these molecules and guide experiments, we must rely on computational means to functionally annotate the majority of sequence data. To understand how well these algorithms perform, we have established a challenge involving a broad scientific community in which we evaluate different annotation methods according to their ability to predict the associations between previously unannotated protein sequences and Gene Ontology terms. Here we discuss the rationale, benefits, and issues associated with evaluating computational methods in an ongoing community-wide challenge.

Key words Function prediction, Algorithms, Evaluation, Machine learning

1 Introduction

Molecular biology has become a high volume information science. This rapid transformation has taken place over the past two decades and has been chiefly enabled by two technological advances: (1) affordable and accessible high-throughput sequencing platforms, sequence diagnostic platforms, and proteomic platforms and (2) affordable and accessible computing platforms for managing and analyzing these data. It is estimated that sequence data accumulates at the rate of 100 exabases per day (1 exabase $=10^{18}$ bases) [35]. However, the available sequence data are of limited use without understanding their biological implications. Therefore, the development of computational methods that provide clues about functional roles of biological macromolecules is of primary importance.

Many function prediction methods have been developed over the past two decades [12, 31]. Some are based on sequence alignments to proteins for which the function has been experimentally established [4, 11, 24], yet others exploit other types of data such as protein structure [26, 27], protein and gene expression data [17], macromolecular interactions [21, 25], scientific literature [3], or a combination of several data types [9, 34, 36]. Typically, each new method is trained and evaluated on different data. Therefore, establishing best practices in method development and evaluating the accuracy of these methods in a standardized and unbiased setting is important. To help choose an appropriate method for a particular task, scientists often form community challenges for evaluating methods [7]. The scope of these challenges extends beyond testing methods: they have been successful in invigorating their respective fields of research by building communities and producing new ideas and collaborations (e.g., [20]).

In this chapter we discuss a community-wide effort whose goal is to help understand the state of affairs in computational protein function prediction and drive the field forward. We are holding a series of challenges which we named the Critical Assessment of Functional Annotation, or CAFA. CAFA was first held in 2010–2011 (CAFA1) and included 23 groups from 14 countries who entered 54 computational function prediction methods that were assessed for their accuracy. To the best of our knowledge, this was the first large-scale effort to provide insights into the strengths and weaknesses of protein function prediction software in the bioinformatics community. CAFA2 was held in 2013–2014, and more than doubled the number of groups (56) and participating methods (126). Although several repetitions of the CAFA challenge would likely give accurate trajectory of the field, there are valuable lessons already learned from the two CAFA efforts.

For further reading on CAFA1, the results were reported in full in [30]. As of this time, the results of CAFA2 are still unpublished and will be reported in the near future. The preprint of the paper is available on arXiv [19].

2 Organization of the CAFA Challenge

We begin our explanation of CAFA by describing the participants. The CAFA challenge generally involves the following groups: the organizers, the assessors, the biocurators, the steering committee, and the predictors (Fig. 1a).

The main role of the organizers is to run CAFA smoothly and efficiently. They advertise the challenge to recruit predictors, coordinate activities with the assessors, report to the steering committee, establish the set of challenges and types of evaluation, and run the CAFA web site and social networks. The organizers also compile CAFA data and coordinate the publication process. The assessors develop assessment rules, write and maintain assessment software, collect the submitted prediction data, assess the data, and present the evaluations to the community. The assessors work together with the organizers and the steering committee on standardizing

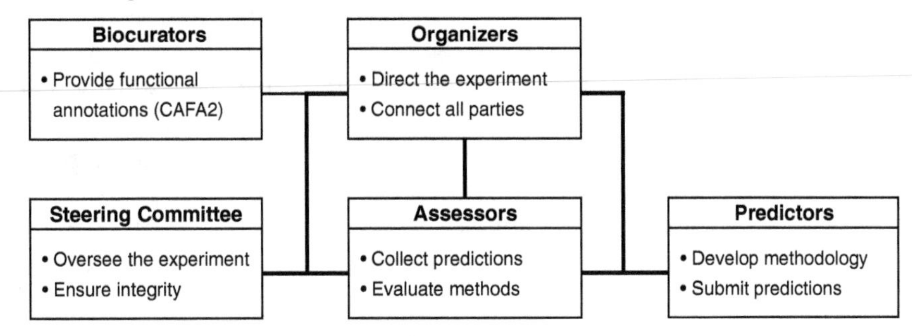

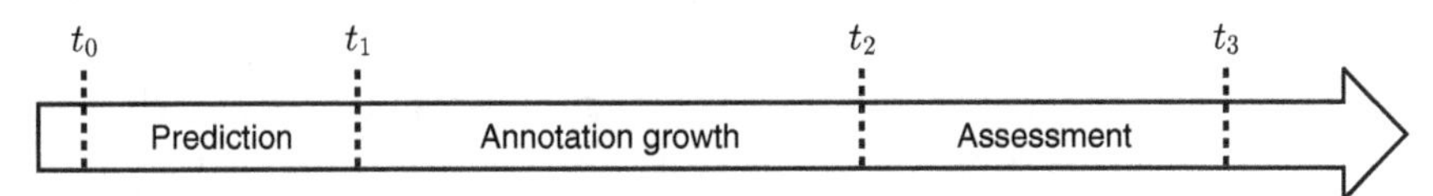

Fig. 1 The organizational structure of the CAFA experiment. (**a**) Five groups of participants in the experiment together with their main roles. Organizers, assessors, and biocurators cannot participate as predictors. (**b**) Timeline of the experiment

submission formats and developing assessment rules. The biocurators joined the experiment during CAFA2: they provide additional functional annotations that may be particularly interesting for the challenge. The steering committee members are in regular contact with the organizers and assessors. They provide advice and guidance that ensures the quality and integrity of the experiment. Finally, the largest group, the predictors, consists of research groups who develop methods for protein function prediction and submit their predictions for evaluation. The organizers, assessors, and biocurators are not allowed to officially evaluate their own methods in CAFA.

CAFA is run as a timed challenge (Fig. 1b). At time t_0, a large number of experimentally unannotated proteins are made public by the organizers and the predictors are given several months, until time t_1, to upload their predictions to the CAFA server. At time t_1 the experiment enters a waiting period of at least several months, during which the experimental annotations are allowed to accumulate in databases such as Swiss-Prot [2] and UniProt-GOA [16]. These newly accumulated annotations are collected at time t_2 and are expected to provide experimental annotations for a subset of original proteins. The performance of participating methods is then analyzed between time points t_2 and t_3 and presented to the community at time t_3. It is important to mention that unlike some machine learning challenges, CAFA organizers do not provide training data that is required to be used. CAFA, thus, evaluates a combination of biological knowledge, the ability to collect and curate training data, and the ability to develop advanced computational methodology.

We have previously described some of the principles that guide us in organizing CAFA [13]. It is important to mention that CAFA is associated with the Automated Function Prediction Special Interest Group (Function-SIG) that is regularly held at the Intelligent Systems for Molecular Biology (ISMB) conference [37]. These meetings provide a forum for exchanging ideas and communicating research among the participants. Function-SIG also serves as the venue at which CAFA results are initially presented and where the feedback from the community is sought.

3 The Gene Ontology Provides the Functional Repertoire for CAFA

Computational function prediction methods have been reviewed extensively [12, 31] and are also discussed in Chapter 5 [8]. Briefly, a function prediction method can be described as a classifier: an algorithm that is tasked with correctly assigning biological function to a given protein. This task, however, is arbitrarily difficult unless the function comes from a finite, preferably small, set of functional terms. Thus, given an unannotated protein sequence and a set of available functional terms, a predictor is tasked with associating terms to a protein, giving a score (ideally, a probability) to each association.

The Gene Ontology (GO) [1] is a natural choice when looking for a standardized, controlled vocabulary for functional annotation. GO's high adoption rate in the protein annotation community helped ensure CAFA's attractiveness, as many groups were already developing function prediction methods based on GO, or could migrate their methods to GO as the ontology of choice. A second consideration is GO's ongoing maintenance: GO is continuously maintained by the Gene Ontology Consortium, edited and expanded based on ongoing discoveries related to the function of biological macromolecules.

One useful characteristic of the basic GO is that its directed acyclic graph structure can be used to quantify the information provided by the annotation; for details on the GO structure see Chaps. 1 and 3 [14, 15]. Intuitively, this can be explained as follows: the annotation term "Nucleic acid binding" is less specific than "DNA binding" and, therefore, is less informative (or has a lower *information content*). (A more precise definition of information content and its use in GO can be found in [23, 32].) The following question arises: if we know that the protein is annotated with the term "Nucleic acid binding," how can we quantify the additional information provided by the term "DNA binding" or incorrect information provided by the term "RNA binding"? The hierarchical nature of GO is therefore important in determining proper metrics for annotation accuracy. The way this is done will be discussed in Sect. 4.2.

When annotating a protein with one or more GO terms, the association of each GO term with the protein should be described using an Evidence Code (EC), indicating how the annotation is supported. For example, the Experimental Evidence code (EXP) is used in an annotation to indicate that an experimental assay has been located in the literature, whose results indicate a gene product's function. Other experimental evidence codes include Inferred by Expression Pattern (IEP), Inferred from Genetic Interaction (IGI), and Inferred from Direct Assay (IDA), among others. Computational evidence codes include lines of evidence that were generated by computational analysis, such as orthology (ISO), genomic context (IGC), or identification of key residues (IKR). Evidence codes are not intended to be a measure of trust in the annotation, but rather a measure of provenance for the annotation itself. However, annotations with experimental evidence are regarded as more reliable than computational ones, having a provenance stemming from experimental verification. In CAFA, we treat proteins annotated with experimental evidence codes as a "gold standard" for the purpose of assessing predictions, as explained in the next section. The computational evidence codes are treated as predictions.

From the point of view of a computational challenge, it is important to emphasize that the hierarchical nature of the GO graph leads to the property of consistency or True Path Rule in functional annotation. Consistency means that when annotating a protein with a given GO term, it is automatically annotated with all the ancestors of that term. For example, a valid prediction cannot include "DNA binding" but exclude "Nucleic acid binding" from the ontology because DNA binding implies nucleic acid binding. We say that a prediction is not consistent if it includes a child term, but excludes its parent. In fact, the UniProt resource and other databases do not even list these parent terms from a protein's experimental annotation. If a protein is annotated with several terms, a valid complete annotation will automatically include all parent terms of the given terms, propagated to the root(s) of the ontology. The result is that a protein's annotation can be seen as a consistent sub-graph of GO. Since any computational method effectively chooses one of a vast number of possible consistent sub-graphs as its prediction, the sheer size of the functional repertoire suggests that function prediction is non-trivial.

4 Comparing the Performance of Prediction Methods

In the CAFA challenge, we ask the participants to associate a large number of proteins with GO terms and provide a probability score for each such association. Having associated a set of GO sub-graphs with a given confidence, the next step is to assess how accurate these predictions are. This involves: (1) establishing standards of truth and (2) establishing a set of assessment metrics.

4.1 Establishing Standards of Truth

The main challenge to establishing a standard-of-truth set for testing function prediction methods is to find a large set of correctly annotated proteins whose functions were, until recently, unknown.

An obvious choice would be to ask experimental scientists to provide these data from their labs. However, scientists prefer to keep the time between discovery and publication as brief as possible, which means that there is only a small window in which new experimental annotations are not widely known and can be used for assessment. Furthermore, each experimental group has its own "data sequestration window" making it hard to establish a common time for all data providers to sequester their data. Finally, to establish a good statistical baseline for assessing prediction method performance, a large number of prediction targets are needed, which is problematic since most laboratories research one or only a few proteins each. High-throughput experiments, on the other hand, provide a large number of annotations, but those tend to concentrate only on few functions, and generally provide annotations that have a lower information content [32].

Given these constraints, we decided that CAFA would not initially rely on direct communication between the CAFA organizers and experimental scientists to provide new functional data. Instead, CAFA relies primarily on established biocuration activities around the world: we use annotation databases to conduct CAFA as a time-based challenge. To do so, we exploit the following dynamics that occurs in annotation databases: protein annotation databases grow over time. Many proteins that at a given time t_1 do not have experimentally verified annotation, but later, some of proteins may gain experimental annotations, as biocurators add these data into the databases. This subset of proteins that were not experimentally annotated at t_1, but gained experimental annotations at t_2, are the ones that we use as a test set during assessment (Fig. 1b). In CAFA1 we reviewed the growth of Swiss-Prot over time and chose 50,000 *target proteins* that had no experimental annotation in the Molecular Function or Biological Process ontologies of GO. At t_2, out of those 50,000 targets we identified 866 *benchmark proteins*; i.e., targets that gained experimental annotation in the Molecular Function and/or Biological Process ontologies. While a benchmark set of 866 proteins constitutes only 1.7% of the number of original targets, it is a large enough set for assessing performance of prediction methods. To conclude, exploiting the history of the Swiss-Prot database enabled its use as the source for standard-of-truth data for CAFA. In CAFA2, we have also considered experimental annotations from UniProt-GOA [16] and established 3681 benchmark proteins out of 100,000 targets (3.7%).

One criticism of a time-based challenge is that when assessing predictions, we still may not have a full knowledge of a protein's function. A protein may have gained experimental validation for function f_1, but it may also have another function, say f_2, associated with it, which has not been experimentally validated by the time t_2. A method predicting f_2 may be judged to have made a false-positive prediction, even though it is correct (only we do not know it yet). This problem, known as the "incomplete knowledge problem" or the "open world problem" [10] is discussed in detail in Chapter 8 [33]. Although the incomplete knowledge problem may impact the accuracy of time-based evaluations, its actual impact in CAFA has not been substantial. There are several reasons for this and are also discussed in, including the robustness of the evaluation metrics used in CAFA, and that the newly added terms may be unexpected and more difficult to predict. The influence of incomplete data and conditions under which it can affect a time-based challenge were investigated and discussed in [18]. Another criticism of CAFA is that the experimental functional annotations are not unbiased because some terms have a much higher frequency than others due to artificial considerations. There are two chief reasons for this bias: first, high-throughput assays typically assign shallow terms to proteins, but being high throughput means they can dominate the experimentally verified annotations in the databases. Second, biomedical research is driven by interest in specific areas of human health, resulting in over-representation of health-related functions [32]. Unfortunately, CAFA1 and CAFA2 could not guarantee unbiased evaluation. However, we will expand the challenge in CAFA3 to collect genome-wide experimental evidence for several biological terms. Such an assessment will result in unbiased evaluation on those specific terms.

4.2 Assessment Metrics

When assessing the prediction quality of different methods, two questions come to mind. First, what makes a good prediction? Second, how can one score and rank prediction methods? There is no simple answer to either of these questions. As GO comprises three ontologies that deal with different aspects of biological function, different methods should be ranked separately with respect to how well they perform in Molecular Function, Biological Process, or the Cellular Component ontologies. Some methods are trained to predict only for a subset of any given GO graph. For example, they may only provide predictions of DNA-binding proteins or of mitochondrial-targeted proteins. Furthermore, some methods are trained only on a single species or a subset of species (say, eukaryotes), or using specific types of data such as protein structure, and it does not make sense to test them on benchmark sets for which they were not trained. To address this issue, CAFA scored methods not only in general performance, but also on specific subsets of proteins taken from humans and model organisms, including *Mus musculus*, *Rattus norvegicus*, *Arabidopsis thaliana*, *Drosophila melanogaster*, *Caenorhabditis elegans*, *Saccharomyces cerevisiae*, *Dictyostelium discoideum*, and *Escherichia coli*. In CAFA2, we extended this evaluation to also assess the methods only on benchmark proteins on which they made predictions; i.e., the methods were not penalized for omitting any benchmark protein.

One way to view function prediction is as an information retrieval problem, where the most relevant functional terms should be correctly retrieved from GO and properly assigned to the amino acid sequence at hand. Since each term in the ontology implies some or all of its ancestors,[1] a function prediction program's task is to assign the best consistent sub-graph of the ontology to each new protein and output a prediction score for this sub-graph and/or each predicted term. An intuitive scoring mechanism for this type of problem is to treat each term independently and provide the precision–recall curve. We chose this evaluation as our main evaluation in CAFA1 and CAFA2.

Let us provide more detail. Consider a single protein on which evaluation is carried out, but keep in mind that CAFA eventually averages all metrics over the set of benchmark proteins. Let now T be a set of experimentally determined nodes and P a non-empty set of predicted nodes in the ontology for the given protein. Precision (pr) and recall (rc) are defined as

$$\mathrm{pr}(P,T) = \frac{|P \cap T|}{|P|}; \qquad \mathrm{rc}(P,T) = \frac{|P \cap T|}{|T|},$$

where $|P|$ is the number of predicted terms, $|T|$ is the number of experimentally determined terms, and $|P \cap T|$ is the number of terms appearing in both P and T; see Fig. 2 for an illustrative example of this measure. Usually, however, methods will associate scores with each predicted term and then a set of terms P will be established by defining a score threshold t; i.e., all predicted terms with scores greater than t will constitute the set P. By varying the decision threshold $t \in [0,1]$, the precision and recall of each method can be plotted as a curve $(\mathrm{pr}(t), \mathrm{rc}(t))_t$, where one axis is the precision and the other the recall; see Fig. 3 for an illustration of pr–rc curves and [30] for pr–rc curves in CAFA1. To compile the precision–recall information into a single number that would allow easy comparison between methods, we used the maximum harmonic mean of precision and recall anywhere on the curve, or the maximum F_1-measure which we call $F_{\max}$

$$F_{\max} = \max_t \left\{ 2 \times \frac{\mathrm{pr}(t) \times \mathrm{rc}(t)}{\mathrm{pr}(t) + rc(t)} \right\},$$

where we modified pr(t) and rc(t) to reflect the dependency on t. It is worth pointing out that the F-measure used in CAFA places equal emphasis on precision and recall as it is unclear which of the two should be weighted more. One alternative to F_1 would be the use of a combined measure that weighs precision over recall, which reflects the preference of many biologists for few answers with a high fraction of correctly predicted terms (high precision) over many answers with a lower fraction of correct predictions (high recall); the rationale for this tradeoff is illustrated in Fig. 3.

[1] Some types of edges in Gene Ontology violate the transitivity property (consistency assumption), but they are not frequent.

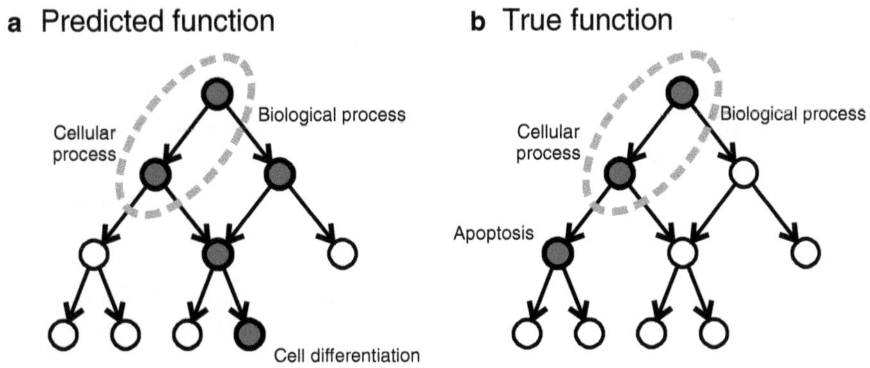

a Predicted function **b** True function

Fig. 2 CAFA assessment metrics. (**a**) *Red nodes* are the predicted terms P for a particular decision threshold in a hypothetical ontology and (**b**) *blue nodes* are the true, experimentally determined terms T. The *circled terms* represent the overlap between the predicted sub-graph and the true sub-graph. There are two nodes (*circled*) in the intersection of P and T, where $|P| = 5$ and $|T| = 3$. This sets the prediction's precision at 2/5=0.4 and recall at 2/3 = 0.667, with $F_1 = 2 \times 0.4 \times 0.667 / (0.4 + 0.667) = 0.5$. The remaining uncertainty (ru) is the information content of the *uncircled blue node* in panel (**b**), while the misinformation (mi) is the total information content of the *uncircled red nodes* in panel (**a**). An information content of any node v is calculated from a representative database as $-\log\Pr(v|\text{Pa}(v))$; i.e., the probability that the node is present in a protein's annotation given that all its parents are also present in its annotation

However, preferring precision over recall in a hierarchical setting can steer methods to focus on shallow (less informative) terms in the ontology and thus be of limited use. At the same time, putting more emphasis on recall may lead to overprediction, a situation in which many or most of the predicted terms are incorrect. For this reason, we decided to equally weight precision and recall. Additional metrics within the precision–recall framework have been considered, though not implemented yet.

Precision and recall are useful because they are easy to interpret: a precision of 1/2 means that one half of all predicted terms are correct, whereas a recall of 1/3 means that one third of the experimental terms have been recovered by the predictor. Unfortunately, precision–recall curves and F_1, while simple and interpretable measures for evaluating ontology-based predictions, are limited because they ignore the hierarchical nature of the ontology and dependencies among terms. They also do not directly capture the information content of the predicted terms. Assessment metrics that take into account the information content of the terms were developed in the past [22, 23, 29], and are also detailed in Chapter 12 [28]. In CAFA2 we used an information-theoretic measure in which each term is assigned a probability that is dependent on the probabilities of its direct parents. These probabilities are calculated from the frequencies of the terms in the database used to generate the CAFA targets. The entire ontology graph, thus, can be seen as a simple Bayesian network [5]. Using this representation, two information-theoretic analogs of precision and recall can be constructed. We refer to these quantities as *misinformation* (mi), the information content attributed to the nodes in the predicted graph that are incorrect, and

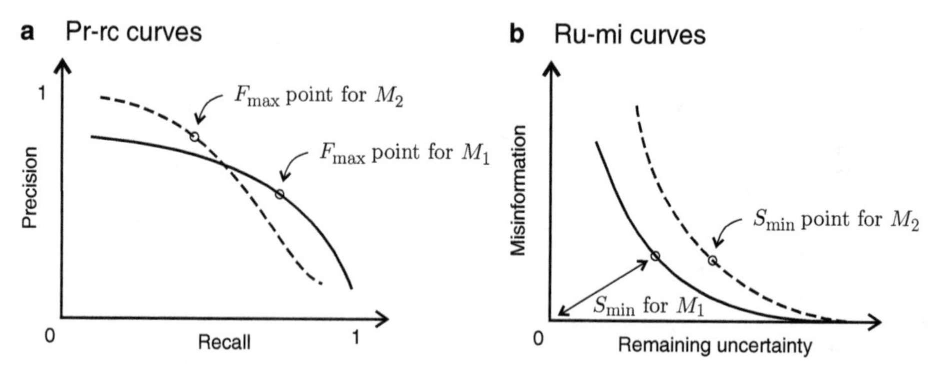

Fig. 3 Precision-recall curves and remaining uncertainty-misinformation curves. This figure illustrates the need for multiple assessment metrics, and understanding the context in which the metrics are used. (**a**) two pr-rc curves corresponding to two prediction methods M_1 and M_2. The point on each curve that gives $F_{\max}$ is marked as a circle. Although the two methods have a similar performance according to $F_{\max}$, method M_1 achieves its best performance at high recall values, whereas method M_2 achieves its best performance at high precision values. (**b**) two ru-mi curves corresponding to the same two prediction methods with marked points where the minimum semantic distance is achieved. Although the two methods have similar performance in the pr-rc space, method M_1 outperforms M_2 in ru-mi space. Note, however, that the performance in ru-mi space depends on the frequencies of occurrence of every term in the database. Thus, two methods may score differently in their $S_{\min}$ when the reference database changes over time, or using a different database

remaining uncertainty (ru), the information content of all nodes that belong to the true annotation but not the predicted annotation. More formally, if T is a set of experimentally determined nodes and P a set of predicted nodes in the ontology, then

$$\mathrm{ru}(P,T) = -\sum_{v \in T-P} \log \Pr(v \mid \mathrm{Pa}(v)); \quad \mathrm{mi}(P,T) = -\sum_{v \in P-T} \log \Pr(v \mid \mathrm{Pa}(v)),$$

where $\mathrm{Pa}(v)$ is the set of parent terms of the node v in the ontology (Fig. 2). A single performance measure to rank methods, the minimum semantic distance $S_{\min}$, is the minimum distance from the origin to the curve $(\mathrm{ru}(t), \mathrm{mi}(t))_t$. It is defined as

$$S_{\min} = \min_t \left\{ (\mathrm{ru}^k(t) + \mathrm{mi}^k(t))^{\frac{1}{k}} \right\},$$

where $k \geq 1$. We typically choose $k = 2$, in which case $S_{\min}$ is the minimum Euclidean distance between the ru–mi curve and the origin of the coordinate system (Fig. 3b). The ru–mi plots and $S_{\min}$ metrics compare the true and predicted annotation graphs by adding an additional weighting component to high-information nodes. In that manner, predictions with a higher information content will be assigned larger weights. The semantic distance has been a useful measure in CAFA2 as it properly accounts for term dependencies in the ontology. However, this approach also has limitations in that it relies on an assumed Bayesian network as a generative model of protein function as well as on the available databases of protein functional annotations where term frequencies change over time. While the latter limitation can be remedied by more robust estimation of term frequencies in a large set of organisms, the performance accuracies in this setting are generally less comparable over two different CAFA experiments than in the precision–recall setting.

5 Discussion

Critical assessment challenges have been successfully adopted in a number of fields due to several factors. First, the recognition that improvements to methods are indeed necessary. Second, the ability of the community to mobilize enough of its members to engage in a challenge. Mobilizing a community is not a trivial task, as groups have their own research priorities and only a limited amount of resources to achieve them, which may deter them from undertaking a time-consuming and competitive effort a challenge may pose. At the same time, there are quite a few incentives to join a community challenge. Testing one's method objectively by a third party can establish credibility, help point out flaws, and suggest improvements. Engaging with other groups may lead to collaborations and other opportunities. Finally, the promise of doing well in a challenge can be a strong incentive heralding a group's excellence in their field. Since the assessment metrics are crucial to the performance of the teams, large efforts are made to create multiple metrics and to describe exactly what they measure. Good challenge organizers try to be attentive to the requests of the participants, and to have the rules of the challenge evolve based on the needs of the community. An understanding that a challenge's ultimate goal is to improve methodologies and that it takes several rounds of repeating the challenge to see results.

The first two CAFA challenges helped clarify that protein function prediction is a vibrant field, but also one of the most challenging tasks in computational biology. For example, CAFA provided evidence that the available function prediction algorithms outperform a straightforward use of sequence alignments in function transfer. The performance of methods in the Molecular Function category has consistently been reliable and also showed progress over time (unpublished results from CAFA2). On the other hand, the performance in the Biological Process or Cellular Component ontologies has not yet met expectations. One of the reasons for this may be that the terms in these ontologies are less predictable using amino acid sequence data and instead would rely more on high-quality systems data; e.g., see [6]. The challenge has also helped clarify the problems of evaluation, both in terms of evaluating over consistent sub-graphs in the ontology but also in the presence of incomplete and biased molecular data. Finally, although it is still early, some best practices in the field are beginning to emerge. Exploiting multiple types of data is typically advantageous, although we have observed that both machine learning expertise and good biological insights tend to result in strong performance. Overall, while the methods in the Molecular Function ontology seem to be maturing, in part because of the strong signal in sequence data, the methods in the Biological Process and Cellular Component ontologies still appear to be in the early stages of development. With the help of better data over time, we expect significant improvements in these categories in the future CAFA experiments.

Overall, CAFA generated a strong positive response to the call for both challenge rounds, with the number of participants substantially growing between CAFA1 (102 participants) and CAFA2 (147). This indicates that there exists significant interest in developing computational protein function prediction methods, in understanding how well they perform, and in improving their performance. In CAFA2 we preserved the experiment rules, ontologies, and metrics we used in CAFA1, but also added new ones to better capture the capabilities of different methods. The CAFA3 experiment will further improve evaluation by facilitating unbiased evaluation for several select functional terms.

More rounds of CAFA are needed to know if computational methods will improve as a direct result of this challenge. But given the community's growth and growing interest, we believe that CAFA is a welcome addition to the community of protein function annotators.

Acknowledgements

We thank Kymberleigh Pagel and Naihui Zhou for helpful discussions. This work was partially supported by NSF grants DBI-1458359 and DBI-1458477. Open Access charges were funded by the University College London Library, the Swiss Institute of Bioinformatics, the Agassiz Foundation, and the Foundation for the University of Lausanne.

References

1. Ashburner M, Ball CA, Blake JA, Botstein D, Butler H, Cherry JM, Davis AP, Dolinski K, Dwight SS, Eppig JT, Harris MA, Hill DP, Issel-Tarver L, Kasarskis A, Lewis S, Matese JC, Richardson JE, Ringwald M, Rubin GM, Sherlock G (2000) Gene ontology: tool for the unification of biology. Nat Genet 25(1):25–29.
2. Bairoch A, Apweiler R, Wu CH, Barker WC, Boeckmann B, Ferro S, Gasteiger E, Huang H, Lopez R, Magrane M, Martin MJ, Natale DA, O'Donovan C, Redaschi N, Yeh LS (2005) The Universal Protein Resource (UniProt). Nucleic Acids Res 33(Database issue):D154–D159
3. Camon EB, Barrell DG, Dimmer EC, Lee V, Magrane M, Maslen J, Binns D, Apweiler R (2005) An evaluation of GO annotation retrieval for BioCreAtIvE and GOA. BMC Bioinformatics 6(Suppl 1):S17
4. Clark WT, Radivojac P (2011) Analysis of protein function and its prediction from amino acid sequence. Proteins 79(7):2086–2096
5. Clark WT, Radivojac P (2013) Information-theoretic evaluation of predicted ontological annotations. Bioinformatics 29(13):i53–i61.

6. Costanzo M, Baryshnikova A, Bellay J, Kim Y, Spear ED, Sevier CS, Ding H, Koh JL, Toufighi K, Mostafavi S, Prinz J, St Onge RP, VanderSluis, B, Makhnevych T, Vizeacoumar FJ, Alizadeh S, Bahr S, Brost RL, Chen Y, Cokol M, Deshpande R, Li Z, Lin ZY, Liang W, Marback M, Paw J, San Luis BJ, Shuteriqi E, Tong AH, van Dyk N, Wallace IM, Whitney JA, Weirauch MT, Zhong G, Zhu H, Houry WA, Brudno M, Ragibizadeh S, Papp B, Pal C, Roth FP, Giaever G, Nislow C, Troyanskaya OG, Bussey H, Bader GD, Gingras AC, Morris QD, Kim PM, Kaiser CA, Myers CL, Andrews BJ, Boone C (2010) The genetic landscape of a cell. Science 327(5964):425–431
7. Costello JC, Stolovitzky G (2013) Seeking the wisdom of crowds through challenge-based competitions in biomedical research. Clin Pharmacol Ther 93(5):396–398
8. Cozzetto D, Jones DT (2016) Computational methods for annotation transfers from sequence. In: Dessimoz C, Škunca N (eds) The gene ontology handbook. Methods in molecular biology, vol 1446. Humana Press. Chapter 5
9. Cozzetto D, Buchan DWA, Bryson K, Jones DT (2013) Protein function prediction by massive integration of evolutionary analyses and multiple data sources. BMC Bioinformatics 14(Suppl 3):S1+.
10. Dessimoz C, Skunca N, Thomas PD (2013) CAFA and the open world of protein function predictions. Trends Genet 29(11):609–610
11. Engelhardt BE, Jordan MI, Muratore KE, Brenner SE (2005) Protein molecular function prediction by Bayesian phylogenomics. PLoS Comput Biol 1(5):e45
12. Friedberg I (2006) Automated protein function prediction–the genomic challenge. Brief Bioinform 7(3):225–242.
13. Friedberg I, Wass MN, Mooney SD, Radivojac P (2015) Ten simple rules for a community computational challenge. PLoS Comput Biol 11(4):e1004150 (2015)
14. Gaudet P, Škunca N, Hu JC, Dessimoz C (2016) Primer on the gene ontology. In: Dessimoz C, Škunca N (eds) The gene ontology handbook. Methods in molecular biology, vol 1446. Humana Press. Chapter 3
15. Hastings J (2016) Primer on ontologies. In: Dessimoz C, Škunca N (eds) The gene ontology handbook. Methods in molecular biology, vol 1446. Humana Press. Chapter 1
16. Huntley RP, Sawford T, Mutowo-Meullenet P, Shypitsyna A, Bonilla C, Martin MJ, O'Donovan C (2015) The GOA database: gene ontology annotation updates for 2015. Nucleic Acids Res 43(Database issue):D1057–D1063
17. Huttenhower C, Hibbs M, Myers C, Troyanskaya OG (2006) A scalable method for integration and functional analysis of multiple microarray datasets. Bioinformatics 22(23):2890–2897
18. Jiang Y, Clark WT, Friedberg I, Radivojac P (2014) The impact of incomplete knowledge on the evaluation of protein function prediction: a structured-output learning perspective. Bioinformatics (Oxford, England) 30(17):i609–i616.
19. Jiang Y, Oron TR, Clark WT, Bankapur AR, D'Andrea D, Lepore R, Funk CS, Kahanda I, Verspoor KM, Ben-Hur A, Koo E, Penfold-Brown D, Shasha D, Youngs N, Bonneau R, Lin A, Sahraeian SME, Martelli PL, Profiti G, Casadio R, Cao R, Zhong Z, Cheng J, Altenhoff A, Skunca N, Dessimoz C, Dogan T, Hakala K, Kaewphan S, Mehryary F, Salakoski T, Ginter F, Fang H, Smithers B, Oates M, Gough J, Toronen P, Koskinen P, Holm L, Chen CT, Hsu WL, Bryson K, Cozzetto D, Minneci F, Jones DT, Chapman S, Dukka BKC, Khan IK, Kihara D, Ofer D, Rappoport N, Stern A, Cibrian-Uhalte E, Denny P, Foulger RE, Hieta R, Legge D, Lovering RC, Magrane M, Melidoni AN, Mutowo-Meullenet P, Pichler K, Shypitsyna A, Li B, Zakeri P, ElShal S, Tranchevent LC, Das S, Dawson NL, Lee D, Lees JG, Sillitoe I, Bhat P, Nepusz T, Romero AE, Sasidharan R, Yang H, Paccanaro A, Gillis J, Sedeno-Cortes AE, Pavlidis P, Feng S, Cejuela JM, Goldberg T, Hamp T, Richter L, Salamov A, Gabaldon T, Marcet-Houben M, Supek F, Gong Q, Ning W, Zhou Y, Tian W, Falda M, Fontana P, Lavezzo E, Toppo S, Ferrari C, Giollo M, Piovesan D, Tosatto S, del Pozo A, Fernández JM, Maietta P, Valencia A, Tress ML, Benso A, Di Carlo S, Politano G, Savino A, Ur Rehman H, Re M, Mesiti M, Valentini G, Bargsten JW, van Dijk ADJ, Gemovic B, Glisic S, Perovic V, Veljkovic V, Veljkovic N, Almeida-e Silva DC, Vencio RZN, Sharan M, Vogel J, Kansakar L, Zhang S, Vucetic S, Wang Z, Sternberg MJE, Wass MN, Huntley RP, Martin MJ, O'Donovan C, Robinson PN, Moreau Y, Tramontano A, Babbitt PC, Brenner SE, Linial M, Orengo CA, Rost B, Greene CS, Mooney SD, Friedberg I, Radivojac P (2016) An expanded evaluation of protein function prediction methods shows an improvement in accuracy. http://arxiv.org/abs/1601.00891
20. Kryshtafovych A, Fidelis K, Moult J (2014) CASP10 results compared to those of previous CASP experiments. Proteins 82: 164–174.
21. Letovsky S, Kasif S (2003) Predicting protein function from protein/protein interaction

data: a probabilistic approach. Bioinformatics 19(Suppl 1):i197–204

22. Lord PW, Stevens RD, Brass A, Goble CA (2003) Investigating semantic similarity measures across the gene ontology: the relationship between sequence and annotation. Bioinformatics 19(10):1275–1283.
23. Lord PW, Stevens RD, Brass A, Goble CA (2003) Semantic similarity measures as tools for exploring the gene ontology. In: Pacific symposium on biocomputing. Pacific symposium on biocomputing, pp 601–612.
24. Martin DM, Berriman M, Barton GJ (2004) GOtcha: a new method for prediction of protein function assessed by the annotation of seven genomes. BMC Bioinformatics 5:178
25. Nabieva E, Jim K, Agarwal A, Chazelle B, Singh M (2005) Whole-proteome prediction of protein function via graph-theoretic analysis of interaction maps. Bioinformatics 21(Suppl 1):i302–i310
26. Pal D, Eisenberg D (2005) Inference of protein function from protein structure. Structure 13(1):121–130 (2005)
27. Pazos F, Sternberg MJ (2004) Automated prediction of protein function and detection of functional sites from structure. Proc Natl Acad Sci USA 101(41):14754–14759
28. Pesquita C (2016) Semantic Similarity in the Gene Ontology. In: Dessimoz C, Škunca N (eds) The gene ontology handbook. Methods in molecular biology, vol 1446. Humana Press. Chapter 12
29. Pesquita C, Faria D, Falcão AO, Lord P, Couto FM (2009) Semantic similarity in biomedical ontologies. PLoS Comput Biol 5(7): e1000443+.
30. Radivojac P, Clark WT, Oron TRR, Schnoes AM, Wittkop T, Sokolov A, Graim K, Funk C, Verspoor K, Ben-Hur A, Pandey G, Yunes JM, Talwalkar AS, Repo S, Souza ML, Piovesan D, Casadio R, Wang Z, Cheng J, Fang H, Gough J, Koskinen P, Törönen P, Nokso-Koivisto J, Holm L, Cozzetto D, Buchan DW, Bryson K, Jones DT, Limaye B, Inamdar H, Datta A, Manjari SK, Joshi R, Chitale M, Kihara D, Lisewski AM, Erdin S, Venner E, Lichtarge O, Rentzsch R, Yang H, Romero AE, Bhat P, Paccanaro A, Hamp T, Kaßner R, Seemayer S, Vicedo E, Schaefer C, Achten D, Auer F, Boehm A, Braun T, Hecht M, Heron M, Hönigschmid P, Hopf TA, Kaufmann S, Kiening M, Krompass D, Landerer C, Mahlich Y, Roos M, Björne J, Salakoski T, Wong A, Shatkay H, Gatzmann F, Sommer I, Wass MN, Sternberg MJ, Škunca N, Supek F, Bošnjak M, Panov P, Džeroski S, Šmuc T, Kourmpetis YA, van Dijk AD, ter Braak CJ, Zhou Y, Gong Q, Dong X, Tian W, Falda M, Fontana P, Lavezzo E, Di Camillo B, Toppo S, Lan L, Djuric N, Guo Y, Vucetic S, Bairoch A, Linial M, Babbitt PC, Brenner SE, Orengo C, Rost B, Mooney SD, Friedberg I (2013) A large-scale evaluation of computational protein function prediction. Nat Methods 10(3):221–227.
31. Rentzsch R, Orengo CA (2009) Protein function prediction–the power of multiplicity. Trends Biotechnol 27(4):210–219.
32. Schnoes AM, Ream DC, Thorman AW, Babbitt PC, Friedberg I (2013) Biases in the experimental annotations of protein function and their effect on our understanding of protein function space. PLoS Comput Biol 9(5):e1003,063+.
33. Škunca N, Roberts RJ, Steffen M (2016) Evaluating computational gene ontology annotations. In: Dessimoz C, Škunca N (eds) The gene ontology handbook. Methods in molecular biology, vol 1446. Humana Press. Chapter 8.
34. Sokolov A, Ben-Hur A (2010) Hierarchical classification of gene ontology terms using the GOstruct method. J Bioinform Comput Biol 8(2):357–376
35. Stephens ZD, Lee SY, Faghri F, Campbell RH, Zhai C, Efron MJ, Iyer R, Schatz MC, Sinha S, Robinson GE (2015) Big data: astronomical or genomical? PLoS Biol 13(7):e1002195+.
36. Troyanskaya OG, Dolinski K, Owen AB, Altman RB, Botstein D (2003) A Bayesian framework for combining heterogeneous data sources for gene function prediction (in *Saccharomyces cerevisiae*). Proc Natl Acad Sci USA 100(14):8348–8353
37. Wass MN, Mooney SD, Linial M, Radivojac P, Friedberg I (2014) The automated function prediction SIG looks back at 2013 and prepares for 2014. Bioinformatics (Oxford, England) 30(14):2091–2092.

10

Evaluating Computational Gene Ontology Annotations

Nives Škunca, Richard J. Roberts, and Martin Steffen

Abstract

Two avenues to understanding gene function are complementary and often overlapping: experimental work and computational prediction. While experimental annotation generally produces high-quality annotations, it is low throughput. Conversely, computational annotations have broad coverage, but the quality of annotations may be variable, and therefore evaluating the quality of computational annotations is a critical concern.

In this chapter, we provide an overview of strategies to evaluate the quality of computational annotations. First, we discuss why evaluating quality in this setting is not trivial. We highlight the various issues that threaten to bias the evaluation of computational annotations, most of which stem from the incompleteness of biological databases. Second, we discuss solutions that address these issues, for example, targeted selection of new experimental annotations and leveraging the existing experimental annotations.

Key words Gene ontology, Evaluation, Tools, Prediction, Annotation, Function

1 Introduction

Sequencing a genome is now routine. However, knowledge of the gene sequence is only the first step toward understanding it; we ultimately want to understand the function(s) of each gene in the cell. Function annotation using computational methods—for example, function propagation via sequence similarity or orthology—can produce high-probability annotations for a majority of gene sequences, the next step toward understanding. But because computational function annotations often generalize the many layers of biological complexity, we are interested in **evaluating** how well these predictions reflect biological reality. In this chapter, we discuss the evaluation of computational predictions.

First, we highlight issues that make the evaluation of computational predictions challenging, with perhaps the primary challenge being the incompleteness of annotation databases: scoring as "wrong" those computational predictions that are not yet proven or disproven could overestimate the count of "incorrect" predictions, and skew perceptions of computational accuracy [1].

Second, we discuss solutions that address various aspects of database incompleteness. For example, some solutions directly address the incompleteness of databases by adding new experimental annotations. Yet another solution leverages existing high-quality annotations in a current release of a database, and retrospectively evaluates previous releases of the annotation databases. Intuitively, those annotations that are unchanged through multiple successive database releases may be expected to be of higher quality. Additional solutions include leveraging negative annotations, though sparse but containing valuable information, or performing extensive experimentation for a subset of functions of interest.

1.1 Sources of Gene Ontology Annotations: Curated and Computational Annotations

In practice, functional annotation of a gene means the assignment of a single label, or a set of labels; for example, this might involve using BLAST to transfer the labels from another gene. A particularly valuable set of labels for denoting gene function are those derived from the controlled vocabulary established by the Gene Ontology (GO) consortium [2], with terms such as "oxygen transporter activity," "hemoglobin complex," and "heme transport," as descriptors of a gene's Molecular Function, Cellular Component, and Biological Process.

But just as important as the annotation label itself is the knowledge of the source of the annotation. Based on their source, there are two main routes to produce annotations in the GO, and the GO Consortium emphasizes this distinction using evidence codes [3], as described in Chap. 3 [4].

The first route of annotating requires curator's expertise when assigning: be it examining primary or secondary literature to assign appropriate annotations, manually examining phylogenetic trees to infer events of function loss and gain, or deciding on sequence similarity thresholds for specific gene families to propagate annotations. As curated annotation is time consuming, the curators streamline their efforts, by focusing annotations on the 12 model organisms ([5] and Fig. 1, left). Consequently, fewer than 1% of proteins have this type of annotation in the UniProt-GOA database. Elsewhere, a recent examination of the annotation of 3.3 million bacterial genes found that fewer than 0.4% of annotations can be documented by experiment, although estimates suggest that the actual number might be above 1% [6].

The second route of annotating, **computational prediction of function**, takes high-quality curated annotations propagates them across proteins in nonmodel organisms. Once the pipeline for the computational prediction has been setup—a task which is by no means trivial—it can be relatively straightforward to obtain computational prediction of function across a large number of biological sequences. Chapter 5 [7] contains a detailed introduction to the methods used in computational annotation.

Computational prediction of function propagates annotations to the vast majority of currently annotated genes (Fig. 1, right).

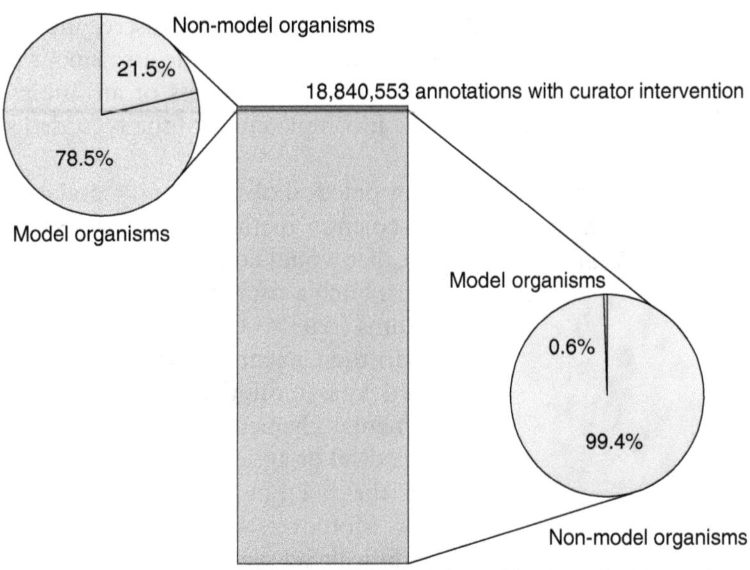

Fig. 1 The distribution of the number of computational annotations obtained *without* curator intervention (evidence code IEA) to all other annotations (evidence codes ISS, IBA, IDA, IMP, ND, IGI, IPI, ISO, TAS, ISA, RCA, IC, NAS, ISM, IEP, IGC, EXP, IRD, IKR). The 12 model organisms are: *Homo sapiens*, *Mus musculus*, *Rattus norvegicus*, *Caenorhabditis elegans*, *Drosophila melanogaster*, *Arabidopsis thaliana*, *Gallus gallus*, *Danio rerio*, *Dictyostelium discoideum*, *Saccharomyces cerevisiae*, *Schizosaccharomyces pombe*, and *Escherichia coli* K-12

Over 99 % of all annotations are created in this manner, and they are applied to approximately 76 % of all genes [6]—the remaining 24 % of genes typically have no annotation or are listed as "hypothetical protein." With the exponential growth of biological databases and the labor-intensive nature of manual curation, it is inevitable that automated computational predictions will provide the vast majority of annotations populating current and future databases.

2 Challenges of Assessing Computational Prediction of Function

Computationally predicted annotations are typically assumed to be less reliable than manually curated ones. Manual curation may be thought of as more cautious, as there is typically a single protein being labeled at a time [8], whereas the goal of computational prediction is typically more ambitious: labeling a large number of proteins—possibly ignoring subtle aspects of the biological reality.

Arguably the most accurate method to evaluate computational predictions of functions is to perform comprehensive experiments (e.g., [9]). However, given the number of computational annotations available, experimental evaluation is prohibitively expensive even for a small subset of the available computational annotations.

As a consequence of this discrepancy in numbers, two practical obstacles interfere with the assessment of computational function prediction: the elusiveness of an unbiased gold standard dataset and the incompleteness of the recorded knowledge.

2.1 The Elusiveness of an Unbiased Gold Standard Dataset

A major practical obstacle to the evaluation of computational function prediction methods is the lack of a gold standard dataset—a dataset that would contain complete annotations for representative proteins. Such a dataset should not be used to train the prediction algorithms (refer to Chap. 5 [7]) and can therefore be used to test them. In the current literature, the validation sets mimic the gold standard dataset, but they are biased:proteins that are prioritized for experimental characterization and curation are often selected for their medical or agricultural relevance, and may not be representative of the full function space that the computational methods address. Moreover, with such incomplete validation sets, it is even more difficult to evaluate algorithms specialized for specific functions—e.g., those identifying membrane-bound proteins. The gold standard dataset needs to cover a large breadth of GO terms and also have comprehensive annotations for these GO terms.

In addition to the difficulties of obtaining a gold standard dataset, the complexity of the GO graph (*see* also Chaps. 14 [10] and 2 [11])—a necessary simplification of the true biological reality—poses obstacles to comparison and evaluation. For example, it is not trivial to compare the prediction scores between the parent (more general) and the child (more specific) GO terms: consider the case when computational methods correctly predict annotations using parent terms, but give erroneous predictions for the child terms, i.e., they overpredict. Alternatively, computational predictions might miss to predict some child GO terms, i.e., they underpredict. One way of handling such situations is to use the structure of the GO to probabilistically model protein function, as described in [12].

2.2 Incomplete Knowledge

Underlying the elusiveness of the unbiased gold standard dataset is the main issue: the incompleteness of the annotation databases. When evaluating computational function annotation methods, we typically compare the predictions with the currently available knowledge. We **confirm** the computational annotation when it is available in our validation set, and we **reject** when its negation is available, e.g., via the NOT qualifier in the GO database. If negative annotations are sparse, as is often the case, it is standard practice to consider wrong a prediction when the predicted annotation is absent from the validation set, e.g., [13]. This is formally called the **Closed World Assumption (CWA)**, the presumption that a statement which is true is also **known** to be true. Conversely, under the CWA, that which is not currently known to be true is considered false.

However, the available knowledge—and consequently the validation set—is incomplete; absence of evidence of function does not imply evidence of absence of function [14]. This is formally referred to as the **Open World Assumption (OWA)**, allowing us to **formalize the concept of incomplete knowledge**. As a consequence of the incompleteness of the validation set, we might be rejecting computational predictions that later prove to be correct [1].

To illustrate the challenges related to the evaluation of function prediction, let us focus on one protein, CLC4E_MOUSE (http://www.uniprot.org/uniprot/Q9R0Q8), in particular to two computational annotations assigned to this protein at the time of writing: the OMA orthology database [15] predicted annotation with "integral component of membrane" (GO:0016021) and the InterPro pipeline predicted annotation with "carbohydrate binding" (GO:0030246). There are no available existing high-quality annotations that confirm these computational predictions.

However, if we take a closer look at these annotations, the OMA annotation "integral component of membrane," compared to the experimental annotation (evidence code IDA) of "receptor activity" is consistent with the experimental annotation: in principle, receptors are integral components of membranes. Additionally, the literature contains evidence that this protein indeed binds carbohydrates [16], thereby confirming the InterPro prediction. Therefore, if we revisit the known annotations and make these statements explicitly known to be true, we can confirm them.

Indeed, for the proteins already present in the UniProt-GOA database, we see that curators do revisited them; more than half of the proteins have already been assigned a new GO term annotation after their first introduction into the database (Fig. 2). An extreme example is provided by the Sonic hedgehog entry in mouse

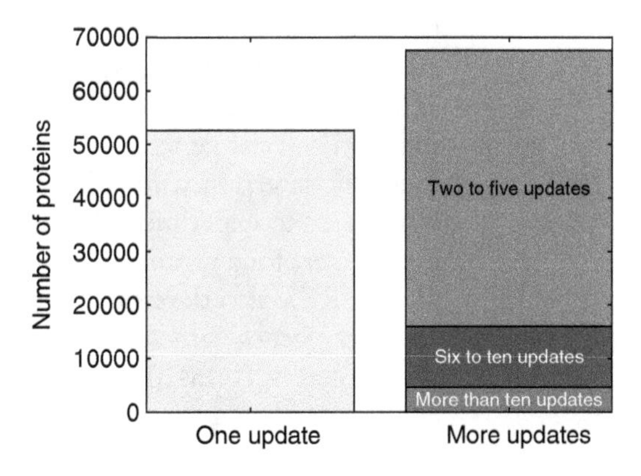

Fig. 2 Distribution of proteins based on the number of times a curator revisits a protein with an annotation from the literature (updates with evidence codes EXP, IDA, IPI, IMP, IGI, IEP). Among the proteins that have a curated annotation based on literature evidence, 56 % are subsequently updated with a new GO term

(http://www.uniprot.org/uniprot/B3GAP8), which has already been revised over a hundred times.

To meaningfully compare computational function annotations, one must account for the Closed World Assumption and have the obstacles it implies in mind. But because of the extent of the gap between the closed and the open world—think of the "unknown unknowns" in the protein function space—a quick-fix solution does not exist. However, numerous ways of tackling the problem were devised, and we turn our attention to those in the subsequent section.

3 Approaches to Test Computational Predictions with Experimental Data

To test computational predictions, experiments have to be conducted. However, the number of proteins that can be experimentally tested are dwarfed by the number of genes identified by genome sequencing, so a very small number of experimental data points must support an enormous number of predicted gene function annotations.

Among the methods to evaluate computational annotations, some are focused on quantifying the available information (e.g., the number and the specificity of annotations) without providing quality judgment (e.g., [17, 18]), while others, the topic of this section, strive to evaluate the quality of the predictions themselves. Addressing some of the complexities of evaluation addressed in the previous section, the latter methods provide good templates for future evaluations of computational methods for function prediction.

3.1 The COMBREX Initiative

The need for experimentally verified annotations is of sufficient scope that it is likely that significant progress can only be made if tackled by the entire scientific community. One such attempt at community building is focused on bacterial proteins: COMBREX (COMputational BRidge to Experiments), along with additional efforts such as the Enzyme Function Initiative [19]. The database (http://combrex.bu.edu) classifies the gene function status of 3.3 million bacterial genes, including 13,665 proteins that have experimentally determined functions [6]. The database contains traceable statements to experimentally characterized proteins, thereby providing support for a given annotation in a clear and transparent manner. COMBREX also developed a tool, named COMBLAST, to associate query genes with the various types of experimental evidence and data stored in COMBREX. COMBLAST output includes a trace to experimental evidence of function via sequence and domain similarity, to available structural information for related proteins, and to association with clinically relevant phenotypes such as antibiotic resistance, and other relevant information. It was used to provide additional annotations for 1474 prokaryotic genomes [20].

Additionally, COMBREX implemented a proof-of-concept prioritization scheme that ranked proteins for experimental testing. For each protein family, distances based on multiple alignments were calculated to help experimentalists easily identify those proteins that might be considered most typical of the family as a whole. The "ideal" COMBREX target is a protein close to many other uncharacterized proteins, and relatively far from any protein of known function, but not so far that it would preclude high-quality predictions of the protein's function for the experimentalist to test.

COMBREX helped fund the implementation of new technology for the experimental characterization of hypothetical proteins from *H. pylori* [21]. A panel of affinity probes was used in a screen to generate initial hypotheses for hypothetical proteins. These hypotheses were then tested and confirmed using traditional in vitro biochemistry. This approach is complementary to other higher throughput methods, such as the parallel screening of metabolite pools [22, 23], and activity-based proteomic approaches to identify proteins of a particular enzymatic class [24, 25].

3.2 CAFA and BioCreAtIvE

CAFA (Critical Assessment of Functional Annotation) is another community-wide effort to evaluate computational annotations, and it promises to uncover some of the most promising algorithms applied to computational function annotation [13]. Such an effort has great utility in establishing success rates of many computational annotation methods based on newly generated curator knowledge. Chapter 10 [26] covers the details of the CAFA evaluation.

Yet another community effort with a more narrow scope, introduced in Chap. 6 [27], BioCreAtIvE (Critical Assessment of Information Extraction systems in Biology) [28] is focused on evaluating annotations obtained through text mining. When evaluating in this setting, the challenges of evaluation within the open/closed world do not exist: methods are evaluated based on the amount of information they can extract from a scientific paper, which in itself has defined bounds. Evaluating the extraction quality of GO annotations for a small set of human proteins showed the extent of the work ahead—text mining algorithms were surpassed by the Precision of expert curators [29]—but also showed the areas that need to be addressed to improve the quality of computational functional annotation using text mining algorithms.

3.3 Evaluating Computational Predictions Over Time Using Successive Database Releases

A strategy to circumvent the problem of the lack of a gold standard is to consider changes in experimental annotations in the UniProt-GOA database [30].

By keeping track of annotations associated with particular proteins across successive releases of the UniProt-GOA database, one can assess the extent to which newly added experimental annotations agree with previous computational predictions. As a surrogate for the intuitive notion of specificity, the authors defined a reliability measure as the ratio of confirmed computational annotations to confirmed and rejected/removed ones. One computational annotation is deemed confirmed or rejected, depending on whether a new, corresponding experimental annotation supports or contradicts it. Furthermore, if a computational annotation is removed, the annotation is deemed implicitly rejected and thus contributes negatively to the reliability measure. As a surrogate for the intuitive notion of sensitivity, coverage was defined as the proportion of newly added experimental annotations that had been correctly predicted by computational annotations in a previous release.

Overall, this work found that electronic annotations are more reliable than generally believed, to an extent that they are competitive with annotations inferred by curators when they use evidence other than experiments from the primary literature. But this work also reported significant variations among inference methods, types of annotations, and organisms. For example, the authors noted an overall high reliability of annotations obtained from mapping Swiss-Prot keywords associated with UniProtKB entries to GO terms. Nevertheless, there were exceptions: GO terms related to metal ion binding had low reliability in the analysis due to a large number of removed annotations. Similarly, a few annotations related to ion transport were explicitly rejected with the 'NOT' qualifier, e.g., for UniProtID Q6R3K9 ('NOT' annotation for "iron ion transport") and UniProtID Q9UN42 ('NOT' annotation for "monovalent inorganic cation transport").

3.4 Increasing the Number of Negative ('NOT') Annotations

Having a comprehensive set of negative annotations would bridge the gap between CWA and OWA; knowing both which functions *are* and are *not* assigned to a protein will not reject predictions that might later prove to be correct.

While experimentally assigning a function to protein is difficult and time consuming, it may be equally challenging to establish that a protein does *not* perform a particular function. For example, unsuccessfully testing a protein for a particular function may only indicate that it is either more difficult to demonstrate such an activity or that it is not present under the given conditions. Because the number and the combination of environmental conditions to test—e.g., the right partners or the right environmental stimulus—is numerous, obtaining a set of 'NOT' annotations might be feasible only for a subset of functions. Consequently, the negative annotations are few and far in between in annotation databases. For example, the January 2015 release of the UniProt-GOA database contains only 8961 entries that are marked with a 'NOT' qualifier.

There is a small number of reports in the literature stating that a protein does not perform a specific function (e.g., [31]), and therefore such sporadic reports cannot be the basis for a comprehensive evaluation of computational annotations. Large-scale production of negative annotations do exists; for example, denoting a set of GO terms that are not likely to be assigned to a protein, given its known annotations (e.g., [32]). However, these are also computational *predictions*, they also need to be evaluated.

3.5 Evaluating Computational Predictions for a Specific Subset of GO Terms

The BioCreAtIvE challenge performed annotations without the challenges of the open and closed world of function annotations by focusing on defined "chunks" of information, scientific papers. In the realm of computational predictions, one of the more straightforward ways of avoiding the challenges of the closed world is to limit the scope to function where we have close to complete comprehension. In fact, by narrowing the scope of the function annotation problem, Huttenhower et al. did just that [9].

The authors evaluated the computational predictions, focusing the evaluation on functions related to mitochondrial organization and biogenesis in *Saccharomyces cerevisiae*. They trained their function prediction models only on the annotation data available in the databases, but performed comprehensive experiments for all genes in *S. cerevisiae* to check whether they have function related to mitochondrial organization and biogenesis. This way, they had information for every *S. cerevisiae* gene and were able to evaluate the prediction accuracy without the need for the distinction between the open and the closed world.

3.6 Simulation Studies

Simulation studies are abundantly used to evaluate computational methods that simulate various evolutionary events, as is done, for example, with the simulation framework for genome evolution Artificial Life Framework (ALF) [33]. In a related application of simulation, simulated erroneous annotations were used to study the quality of computational annotations—curated GO annotations obtained using methods based on sequence similarity, in the GO database denoted with the evidence code ISS [34]. First, the authors estimated the level of errors among the ISS GO annotations by checking for the effect of randomly adding erroneous annotations. Second, they obtained a linear model that connected the propensity of (artificially introduced) errors among the annotations with the estimate of Precision. Finally, they used this model to estimate the baseline Precision at the level where there are no introduced errors.

4 Outlook

Experimental annotations are key to evaluate computational methods to predict annotations. Therefore, it is highly desirable that three principles govern experimental testing of gene function: maximal leveraging of existing experimental information, maximal information gain with each new experiment, and the development of higher throughput approaches.

Maximal leveraging of existing experimental information is easiest to obtain through the use of traceable statements, such as the use of the "with" field in the UniProt-GOA database: the "with" field can record the protein that was used as template to transfer annotation through sequence similarity. However, we could go a step further, toward statements such as: "Gene X has 96.8 % sequence identity to the experimentally characterized protein 'HP0050' and therefore this protein is annotated as 'adenine specific DNA methyltransferase'." Traceable statements greatly increase the transparency of a prediction, and allow the users of gene annotations to estimate their confidence in the annotation, regardless of the source—manual curator or an automated computational prediction [35].

In order to increase information gain of new experiments, it would be beneficial to develop and incorporate experimental design principles that help guide the identification of maximally informative targets for function validation. One way to maximize the information gain from the experimental analysis is to choose proteins that generate or improve predictions for many other proteins across many genomes, as opposed to proteins related to few or no other proteins. Alternatively, for function prediction methods that report probabilities, the information gain from an experiment can be quantified as the reduction in the estimated probability of prediction error, summed across all predictions [36].

Development of higher throughput approaches for the testing of protein function is well underway, and we can hope for the same effects as with DNA sequencing. However, at the time of writing, a small number of experimental studies contribute much of the functional protein annotations collected in the databases, thereby biasing the available experimental annotations [8]. Indeed, DNA sequencing did not achieve its dramatic cost reductions and increases in throughput fortuitously, but rather was the result of the systematic investment of hundreds of millions of dollars in technology development over two decades.

Traditionally, the increases of success rates associated with computational function annotation are attributed to methodological refinements. However, we must also quantify the influence of the data available—e.g., more sequences and more function annotations—*independently* of the influence of the algorithms. This information is critical, if only because of the rate of aggregation of new information in the bioinformatics databases. Indeed, an increase in the number of sequenced genomes and an increase in the number of function annotations has a dramatic positive effect on predictive accuracy of at least one computational method of function annotation, phylogenetic profiling [37].

5 Conclusion

There are a plethora of highly accurate, readily available computational function annotation methods available to scientists, and state-of-the-art computational function annotations, such as in the UniProt-GOA database, are easily accessible to all. However, without transparent evaluation and benchmarking, it is still extremely challenging to differentiate among annotations, and annotation methods.

Going forward, the biocuration community will continue to advance along three important lines: increased amounts of biological sequence to be annotated, increased numbers of high-quality experimental annotations, and increased predictive accuracy of computational methods of annotation. In order to achieve the greatest increase in biological knowledge, we will couple the advances made in each of these three areas to reach other, especially coupling advances in the development of new algorithms with robust evaluations of these algorithms based on experimental data, with the purpose of generating new, useful biological hypotheses. Such work will contribute to closing the gap between the Open and the Closed worlds, and greatly increase our understanding of the large number new sequences that are now generated daily.

Acknowledgments

The authors thank Christophe Dessimoz and Maria Anisimova for helpful comments and suggestions. Open Access charges were funded by the University College London Library, the Swiss Institute of Bioinformatics, the Agassiz Foundation, and the Foundation for the University of Lausanne.

References

1. Dessimoz C, Škunca N, Thomas PD (2013) CAFA and the open world of protein function predictions. Trends Genet 29:609–610
2. Ashburner M, Ball CA, Blake JA et al (2000) Gene ontology: tool for the unification of biology. The Gene Ontology Consortium. Nat Genet 25:25–29
3. Guide to GO Evidence Codes | Gene Ontology Consortium. http://geneontology.org/page/guide-go-evidence-codes.
4. Gaudet P, Škunca N, Hu JC, Dessimoz C (2016) Primer on the gene ontology. In: Dessimoz C, Škunca N (eds) The gene ontology handbook. Methods in molecular biology, vol 1446. Humana Press. Chapter 3
5. Reference Genome Group of the Gene Ontology Consortium (2009) The Gene Ontology's Reference Genome Project: a unified framework for functional annotation across species. PLoS Comput Biol 5:e1000431
6. Anton BP, Chang Y-C, Brown P et al (2013) The COMBREX project: design, methodology, and initial results. PLoS Biol 11:e1001638
7. Cozzetto D, Jones DT (2016) Computational methods for annotation transfers from sequence. In: Dessimoz C, Škunca N (eds) The gene ontology handbook. Methods in molecular biology, vol 1446. Humana Press. Chapter 5
8. Schnoes AM, Ream DC, Thorman AW et al (2013) Biases in the experimental annotations of protein function and their effect on our understanding of protein function space. PLoS Comput Biol 9:e1003063

9. Huttenhower C, Hibbs MA, Myers CL et al (2009) The impact of incomplete knowledge on evaluation: an experimental benchmark for protein function prediction. Bioinformatics 25:2404–2410
10. Gaudet P, Dessimoz C (2016) Gene ontology: pitfalls, biases, and remedies. In: Dessimoz C, Škunca N (eds) The gene ontology handbook. Methods in molecular biology, vol 1446. Humana Press. Chapter 14
11. Thomas PD (2016) The gene ontology and the meaning of biological function. In: Dessimoz C, Škunca N (eds) The gene ontology handbook. Methods in molecular biology, vol 1446. Humana Press. Chapter 2
12. Clark WT, Radivojac P (2013) Information-theoretic evaluation of predicted ontological annotations. Bioinformatics 29:i53–i61
13. Radivojac P, Clark WT, Oron TR et al (2013) A large-scale evaluation of computational protein function prediction. Nat Methods 10:221–227
14. Thomas PD, Wood V, Mungall CJ et al (2012) On the use of gene ontology annotations to assess functional similarity among orthologs and paralogs: a short report. PLoS Comput Biol 8:e1002386
15. Altenhoff AM, Skunca N, Glover N et al (2014) The OMA orthology database in 2015: function predictions, better plant support, synteny view and other improvements. Nucleic Acids Res 43(Database issue):D240–D249
16. Yamasaki S, Matsumoto M, Takeuchi O et al (2009) C-type lectin Mincle is an activating receptor for pathogenic fungus, Malassezia. Proc Natl Acad Sci U S A 106:1897–1902
17. Buza TJ, McCarthy FM, Wang N et al (2008) Gene ontology annotation quality analysis in model eukaryotes. Nucleic Acids Res 36:e12
18. del Pozo A, Pazos F, Valencia A (2008) Defining functional distances over gene ontology. BMC Bioinformatics 9:50
19. Gerlt JA, Allen KN, Almo SC et al (2011) The enzyme function initiative. Biochemistry 50:9950–9962
20. Wood DE, Lin H, Levy-Moonshine A et al (2012) Thousands of missed genes found in bacterial genomes and their analysis with COMBREX. Biol Direct 7:37
21. Choi H-P, Juarez S, Ciordia S et al (2013) Biochemical characterization of hypothetical proteins from Helicobacter pylori. PLoS One 8:e66605
22. Proudfoot M, Kuznetsova E, Sanders SA et al (2008) High throughput screening of purified proteins for enzymatic activity. Methods Mol Biol 426:331–341
23. Kuznetsova E, Proudfoot M, Sanders SA et al (2005) Enzyme genomics: application of general enzymatic screens to discover new enzymes. FEMS Microbiol Rev 29:263–279
24. Cravatt BF, Wright AT, Kozarich JW (2008) Activity-based protein profiling: from enzyme chemistry to proteomic chemistry. Annu Rev Biochem 77:383–414
25. Simon GM, Cravatt BF (2010) Activity-based proteomics of enzyme superfamilies: serine hydrolases as a case study. J Biol Chem 285:11051–11055
26. Friedberg I, Radivojac P (2016) Community-wide evaluation of computational function prediction. In: Dessimoz C, Škunca N (eds) The gene ontology handbook. Methods in molecular biology, vol 1446. Humana Press. Chapter 10
27. Ruch P (2016) Text mining to support gene ontology curation and vice versa. In: Dessimoz C, Škunca N (eds) The gene ontology handbook. Methods in molecular biology, vol 1446. Humana Press. Chapter 6
28. Krallinger M, Morgan A, Smith L et al (2008) Evaluation of text-mining systems for biology: overview of the Second BioCreative community challenge. Genome Biol 9(Suppl 2):S1
29. Camon EB, Barrell DG, Dimmer EC et al (2005) An evaluation of GO annotation retrieval for BioCreAtIvE and GOA. BMC Bioinformatics 6(Suppl 1):S17
30. Skunca N, Altenhoff A, Dessimoz C (2012) Quality of computationally inferred gene ontology annotations. PLoS Comput Biol 8:e1002533
31. Poux S, Magrane M, Arighi CN et al (2014) Expert curation in UniProtKB: a case study on dealing with conflicting and erroneous data. Database:bau016
32. Youngs N, Penfold-Brown D, Bonneau R et al (2014) Negative example selection for protein function prediction: The NoGO Database. PLoS Comput Biol 10:e1003644
33. Dalquen DA, Anisimova M, Gonnet GH et al (2012) ALF—a simulation framework for genome evolution. Mol Biol Evol 29:1115–1123
34. Jones CE, Brown AL, Baumann U (2007) Estimating the annotation error rate of curated GO database sequence annotations. BMC Bioinformatics 8:170
35. Bastian FB, Chibucos MC, Gaudet P et al (2015) The Confidence Information Ontology: a step towards a standard for asserting confidence in annotations. Database:bav043

36. Letovsky S, Kasif S (2003) Predicting protein function from protein/protein interaction data: a probabilistic approach. Bioinformatics 19(Suppl 1):i197–i204

37. Škunca N, Dessimoz C (2015) Phylogenetic profiling: how much input data is enough? PLoS One 10:e0114701

Using the Gene Ontology

11

A Gene Ontology Tutorial in Python

Alex Warwick Vesztrocy and Christophe Dessimoz

Abstract

This chapter is a tutorial on using Gene Ontology resources in the Python programming language. This entails querying the Gene Ontology graph, retrieving Gene Ontology annotations, performing gene enrichment analyses, and computing basic semantic similarity between GO terms. An interactive version of the tutorial, including solutions, is available at http://gohandbook.org.

Key words Gene Ontology, Tutorial, Python

1 Introduction

One of the main goals of developing a formal ontology is to facilitate computational analysis. The purpose of this chapter is to provide a hands-on introduction to handling GO terms and GO annotations in Python. This tutorial also shows how Python can be used to perform GO term enrichment analyses, as well as how to compute the similarity between GO terms.

This tutorial uses Python, but other popular languages commonly used to perform GO analyses include Java, R, Perl, and Matlab. The Gene Ontology consortium website maintains a list of software libraries, accessible from

ftp://ftp.geneontology.org/pub/go/www/GO.tools_by_type.software.shtml

An interactive version of this tutorial, with model solutions to all the questions, is available from the book homepage at http://gohandbook.org.

2 Querying the Gene Ontology

A fundamental first step is to retrieve the Gene Ontology and analyse that structure (Chap. 3 [1]).

One convenient Python package available to query the GO is GOATOOLS [2]. This package can read the GO structure stored in OBO format, which is available from the GO website (see Chap. 11 [3]). After loading this file, it is possible to traverse the GO structure, search for particular GO terms, and find out which other terms they are related to and how.

This package is available on the Python Package Index (PyPI), a standard repository of python libraries. As such, it is possible to install it locally using the command[1]:

```
pip install goatools
```

The GOATOOLS package contains the functions necessary to parse the GO in OBO format, to query it, and to visualise the ontology. Using the function `obo_parser.GODag()` from GOATOOLS, the GO file can be loaded. Each GO term in the resulting object is an instance of the `GOTerm` class, which contains many useful attributes, such as:

- `GOTerm.name`: textual definition;
- `GOTerm.namespace`: the ontology the term belongs to (i.e., Molecular Function [MF], Biological Process [BP], or Cellular Component [CC]);
- `GOTerm.parents`: list of parent terms;
- `GOTerm.children`: list of children terms;
- `GOTerm.level`: shortest distance to the root node;

Exercise 2.1

Download the GO basic file in OBO format (`go-basic.obo`), and load the GO using the function `obo_parser.GODag()` from GOATOOLS. Using this library, answer the following questions:

(a) What is the name of the GO term `GO:0048527`?

(b) What are the immediate parent(s) of the term `GO:0048527`?

(c) What are the immediate children of the term `GO:0048527`?

(d) Recursively find all the parent and child terms of the term `GO:0048527`. *Hint*: use your solutions to the previous two questions, with a recursive loop.

(e) How many GO terms have the word "*growth*" in their name?

(f) What is the deepest common ancestor term of `GO:0048527 and GO:0097178`?

(g) Which GO terms regulate `GO:0007124` (pseudohyphal growth)? *Hint*: load the relationship tags and look for terms which define regulation.

[1] GOATOOLS version 0.6.4 was used to write this tutorial and the exercises. To install this exact version, use `pip install goatools==0.6.4`

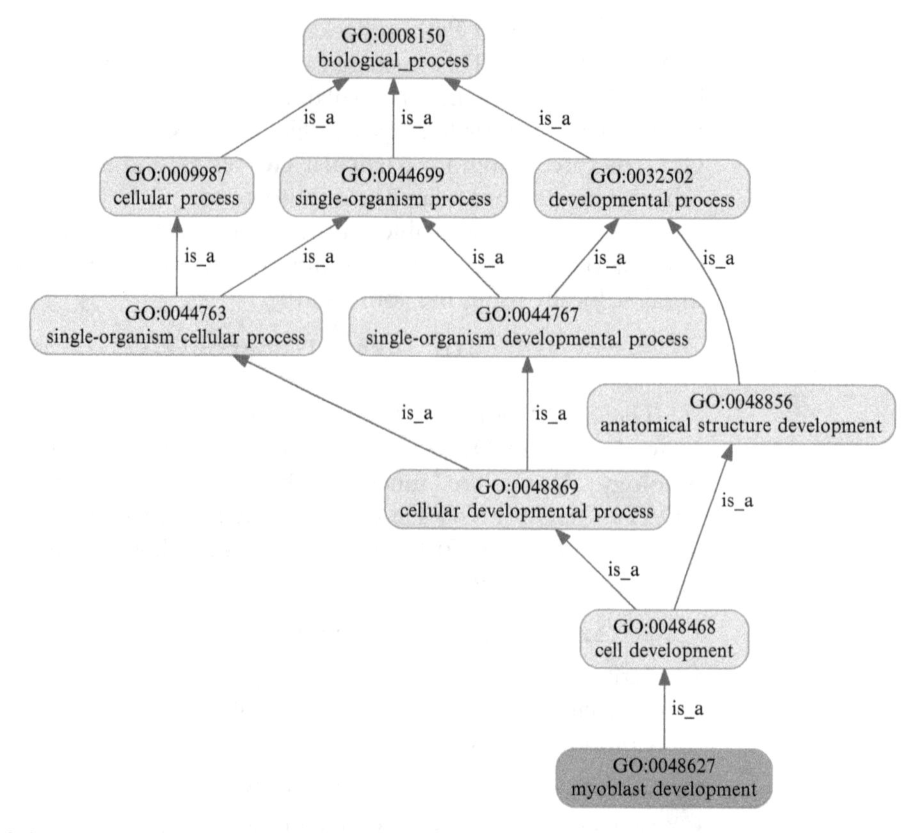

Fig. 1 Selected parts of the Gene Ontology can be visualised using the GOATOOLS library [2]

Exercise 2.2
Using the visualisation function in the GOATOOLS library, answer the following questions:

(a) Produce a figure similar to that in Fig. 1, for the GO term `GO:0097190`. From the visualisation, what is the name of this term?

(b) Using this figure, what is the most specific term that is in the parent terms of both `GO:0097191` (extrinsic apoptotic signalling pathway) and `GO:0038034` (signal transduction in absence of ligand)? This is also referred to as the lowest common ancestor (see Chap. 12 [4]).

Furthermore, other tag-value lines such as the “relationships” can be loaded with an optional argument of, e.g., `optional_attrs=['relationship']`.

The GOATOOLS library also includes functions to visualise the GO graph. For instance, it is possible to depict the location of a particular GO term in the ontology using the method `GOTerm.draw_lineage()`. For example, the plot in Fig. 1 showing the lineage of the GO term `GO:0048527` was created using this function.

As an alternative to GOATOOLS and OBO files, it is possible to retrieve information relating to a specific term from a web service. One such service is the EMBL-EBI QuickGO resource (see

Chap. 11; [3, 5]), which can provide descriptive information about GO terms in OBO-XML format. It is possible to request this OBO-XML file over HTTP, using a URL of the form

http://www.ebi.ac.uk/QuickGO/GTerm?id=<GO_ID>&
format=oboxml

where `<GO_ID>` is replaced with the GO identifier for the term of interest. In Source Code 2.1, an example function to automate this in Python is listed, which uses the urllib library to request the OBO-XML and the xmltodict library to parse the XML into an easy to use dictionary structure. Both libraries are available to install using `pip`, if required. Note that the future library was used to ensure that the function is both Python 2 and 3 compatible.

The dictionary structure that is returned can vary based on what information is available in the database. One example of an information-rich term is `GO:0043065`. A visualisation of the dictionary

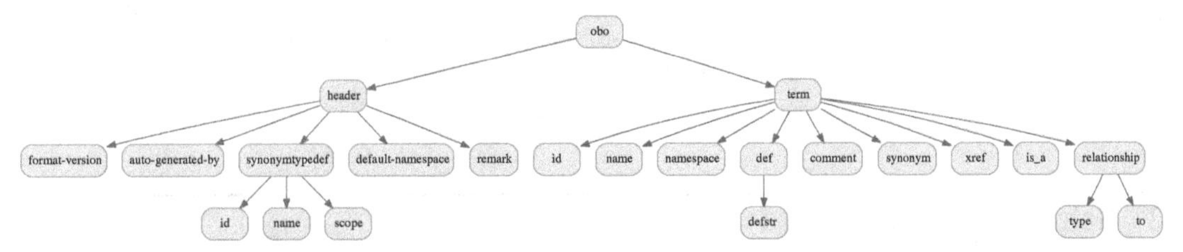

Fig. 2 Visualisation of the keys in the hierarchical dictionary structure returned by `get_oboxml('GO:0043065')`

Source Code 2.1. `get_oboxml()` function for Python 2 and 3.

```
from future.standard_library import install_aliases
install_aliases()
from urllib.request import urlopen
import xmltodict

def get_oboxml(go_id):
    """
        This function retrieves the OBO-XML for a
        given Gene Ontology term, using EMBL-EBI's
        QuickGO browser.
        Input: go_id - a valid Gene Ontology ID,
        e.g. GO:0048527.
    """
    quickgo_url= "http://ebi.ac.uk/QuickGO/GTerm?id="+
    go_id+"&format=oboxml"
    oboxml = urlopen(quickgo_url)

    # Check the response
    if(oboxml.getcode() == 200):
        obodict = xmltodict.parse(oboxml.read())
        return obodict
    else:
        raise ValueError("Couldn't receive OBOXML
        from QuickGO. Check URL and try again.")
```

structure for this term, created with the `visualisedictionary` package available from PyPI (using `pip`), has been included in Fig. 2.

The main advantage of using a web service, such as QuickGO, is that there is no requirement to download and parse the entire Gene Ontology structure; only the information required is retrieved. This is therefore more efficient if only a few particular terms are involved in an analysis. By contrast, for analyses involving many terms, the file-based approach described above is more suitable.

Exercise 2.3
Using the function `get_oboxml()`, listed in Source Code 2.1, answer the following questions:

(a) Find the name and description of the GO term `GO:0048527` (lateral root development). *Hint*: print out the dictionary returned by the function and study its structure, or use the visualisation in Fig. 2.

(b) Look at the difference in the OBO-XML output for the GO terms `GO:00048527` (lateral root development) and `GO:0097178` (ruffle assembly), then generate a table of the synonymous relationships of the term `GO:0097178`.

3 Retrieving GO Annotations

This section looks at manipulating the Gene Association File (GAF) standard, using a parser from the BioPython package [6].

Firstly, a GAF file, which contains GO annotations, shall be downloaded from the UniProt-GOA database [7]. Their website (https://www.ebi.ac.uk/GOA/downloads) lists a number of variants. For this tutorial the reduced GAF file containing only the gene association data for *Arabidopsis thaliana* is going to be used.

Annotations from GAF files can be loaded into a Python dictionary using an iterator from the BioPython package (`Bio.UniProt.GOA.gafiterator`). Source Code 3.1 shows a simple example of this being used, in order to print out the protein ID for each annotation.

Source Code 3.1

```
from Bio.UniProt.GOA import gafiterator
import gzip

# filename = <LOCATION OF GAF FILE>
filename = 'gene_association.goa_arabidopsis.gz'

with gzip.open(filename, 'rt') as fp:
    for annotation in gafiterator(fp):
        # Output annotated protein ID
        print(annotation['DB_Object_ID'])
```

Recall that the latest GAF standard, version 2.1, has 17 tab-delimited fields, which are described in detail in Chap. 3 [1]. Some of them include:

- `'DB'`: the protein database;
- `'DB_Object_ID'`: protein ID;
- `'Qualifier'`: annotation qualifier (such as NOT);
- `'GO_ID'`: GO term;
- `'Evidence'`: evidence code.

Exercise 3.1

(a) Find the total number of annotations for *Arabidopsis thaliana* with NOT qualifiers. What is this as a percentage of the total number of annotations for this species?

(b) How many genes (of *Arabidopsis thaliana*) have the annotation `GO:0048527` (lateral root development)?

(c) Generate a list of annotated proteins which have the word "growth" in their name.

(d) There are 21 evidence codes used in the Gene Ontology project. As discussed in Chap. 3 [1], many of these are inferred, either by curators or automatically. Find the counts of each evidence code in the *Arabidopsis thaliana* annotation file.

4 GO Enrichment or Depletion Analysis

As discussed in detail in Chap. 13 [8] one of the most common analyses performed on GO data is an enrichment (or depletion) analysis. In this tutorial, the `GOEnrichmentStudy()` function available in the GOATOOLS library (which has been seen in section 2) will be used.

The `GOEnrichmentStudy()` function requires the following arguments:

1. the background set of terms (also known as the "population set"), passed as a list of GO term IDs;
2. associations between proteins IDs and GO term IDs, passed as a dictionary with protein IDs as the keys and sets of associated GO terms as the values;
3. the Gene Ontology structure, i.e., the output by the `obo_parser()` function from GOATOOLS;
4. whether annotations should be propagated to all parent terms, (defined in terms of `is_a` tags, only), indicated by setting the optional boolean parameter `propagate_counts` to `True` (default) or `False`;

5. the significance level, indicated by setting the optional parameter `alpha` to the desired cut-off (default: 0.05);
6. the foreground set of terms (also known as "study set"), indicated by setting the parameter `study` to a list of GO term IDs;
7. the list of method(s) to be used to assess significance, indicated by setting the parameter `methods` to a list containing one or several of these elements:
   (a) `"bonferroni"`: Fisher's exact test with Bonferroni correction for multiple testing;
   (b) `"sidak"`: Fisher's exact test with Šidák correction for multiple testing;
   (c) `"holm"`: Fisher's exact test with Holm–Bonferroni correction for multiple testing;
   (d) `"fdr"`: Fisher's exact test, controlling the false discovery rate (see Chap. 13 [8]).

The function returns the list of over-represented and under-represented GO terms in the population set, compared to the background set.

Exercise 4.1
Perform an enrichment analysis using the list of genes with the "growth" keyword from exercise 3.1.c. Use the *Arabidopsis thaliana* annotation set as background, also from exercise 3.1, and the GO structure from exercise 2.1.

(a) Which GO term is most significantly enriched or depleted? Does this make sense?
(b) How many terms are enriched, when using the Bonferroni corrected p-value ≤ 0.01?
(c) How many terms are enriched, when using the false discovery rate (a.k.a. q-value) ≤ 0.01?

5 Computing Basic Semantic Similarities Between GO Terms

In this section, the focus is on computing semantic similarity between GO terms, based on ideas presented in detail in Chap. 12 [4]. Semantic similarity measures enable us to quantify the functional similarity of genes annotated with GO terms.

Recall that semantic similarity measures are broadly separated in two categories: graph-based and information-theoretic measures. The former relies only on the structure of the Gene Ontology graph, whilst the latter also accounts for the information content of the terms.

One graph-based measure of semantic similarity, presented in Chap. 12 [4], is the inverse of the number of edges separating two terms. It is possible to compute the minimum number of edges separating two terms (t_1, t_2) by first finding the deepest common ancestor (t_{DCA}). Then the difference in depth between each term and the deepest common ancestor can be used to calculate the minimum distance between the terms. i.e.,

$$\text{min\_distance}(t_1, t_2) = \text{depth}(t_1) + \text{depth}(t_2) - 2 \times \text{depth}(t_{DCA})$$

Further, one example of an information-theoretic measure (see Chap. 12 [4]) is Resnik's similarity measure—the information content of the most informative common ancestor of the two terms in question. The information content of a term is defined as the negative logarithm of its probability, which can be estimated from the frequency of the term in the annotation database of choice.

Exercise 5.1

(a) `GO:0048364` (root development) and `GO:0044707` (single-multicellular organism process) are two GO terms taken from Fig. 1. Calculate the semantic similarity between them based on the inverse of the semantic distance (number of branches separating them).

(b) Calculate the information content (IC) of the GO term `GO:0048364` (root development), based on the frequency of observation in *Arabidopsis thaliana*.

(c) Calculate the Resnik similarity measure between the same two terms as in part a.

Acknowledgements

We thank Adrian Altenhoff, Debra Klopfenstein, and Haibao Tang for helpful feedback on the tutorial. CD acknowledges Swiss National Science Foundation grant 150654 and UK BBSRC grant BB/M015009/1. Open Access charges were funded by the University College London Library, the Swiss Institute of Bioinformatics, the Agassiz Foundation, and the Foundation for the University of Lausanne.

References

1. Gaudet P, Škunca N, Hu JC, Dessimoz C (2016) Primer on the gene ontology. In: Dessimoz C, Škunca N (eds) The gene ontology handbook. Methods in molecular biology, vol 1446. Humana Press. Chapter 3
2. Tang H, Klopfenstein D, Pedersen B et al (2015) GOATOOLS: tools for gene ontology, Zenodo
3. Munoz-Torres M, Carbon S (2016) Get GO! retrieving GO data using AmiGO, QuickGO,

API, files, and tools. In: Dessimoz C, Škunca N (eds) The gene ontology handbook. Methods in molecular biology, vol 1446. Humana Press. Chapter 11

4. Pesquita C (2016) Semantic similarity in the gene ontology. In: Dessimoz C, Škunca N (eds) The gene ontology handbook. Methods in molecular biology, vol 1446. Humana Press. Chapter 12
5. Binns D, Dimmer E, Huntley R et al (2009) QuickGO: a web-based tool for Gene Ontology searching. Bioinformatics 25:3045–3046
6. Cock PJA, Antao T, Chang JT et al (2009) Biopython: freely available Python tools for computational molecular biology and bioinformatics. Bioinformatics 25:1422–1423
7. Huntley RP, Sawford T, Mutowo-Meullenet P et al (2015) The GOA database: gene Ontology annotation updates for 2015. Nucleic Acids Res 43:D1057–63
8. Bauer S (2016) Gene-category analysis. In: Dessimoz C, Škunca N (eds) The gene ontology handbook. Methods in molecular biology, vol 1446. Humana Press. Chapter 13

12

Gene Ontology: Pitfalls, Biases and Remedies

Pascale Gaudet and Christophe Dessimoz

Abstract

The Gene Ontology (GO) is a formidable resource, but there are several considerations about it that are essential to understand the data and interpret it correctly. The GO is sufficiently simple that it can be used without deep understanding of its structure or how it is developed, which is both a strength and a weakness. In this chapter, we discuss some common misinterpretations of the ontology and the annotations. A better understanding of the pitfalls and the biases in the GO should help users make the most of this very rich resource. We also review some of the misconceptions and misleading assumptions commonly made about GO, including the effect of data incompleteness, the importance of annotation qualifiers, and the transitivity or lack thereof associated with different ontology relations. We also discuss several biases that can confound aggregate analyses such as gene enrichment analyses. For each of these pitfalls and biases, we suggest remedies and best practices.

Key words Gene ontology, Gene/protein annotation, Data mining, Bias, Confounding, Simpson's paradox

1 Introduction

As we have seen in previous chapters (for example refer to Chap. 1 [1], Chap. 12 [2], Chap. 13 [3]), by providing a large amount of structured information, the Gene Ontology (GO) greatly facilitates large-scale analyses and data mining. A very common type of analysis entails comparing sets of genes in terms of their functional annotations, for instance to identify functions that are enriched or depleted in particular subsets of genes (Chap. 13 [3]) or to assess whether particular aspects of gene function might be associated with other aspects of genes, such as sequence divergence or regulatory networks.

Despite conscious efforts to keep GO data as normalized as possible, it is heterogeneous in many respects—to a large extent simply because the body of knowledge underlying the GO is itself very heterogeneous. This can introduce considerable biases when the data is used in other analysis, an effect that is magnified in large-scale comparisons.

Statisticians and epidemiologists make a clear distinction between *experimental data*—data from a controlled experiment, designed such that the case and control groups are as identical as possible in all respects other than a factor of interest—and *observational data*—data readily available, but with the potential presence of unknown or unmeasured factors that may confound the analysis. GO annotations clearly falls into the second category. Therefore, testing and controlling for potential confounders is of paramount importance.

Before we go through some of the key biases and known potential confounders, let us consider Simpson's Paradox, which provides a stark illustration of the perils of data aggregation.

1.1 Simpson's Paradox: The Perils of Data Aggregation

Simpson's paradox is the counterintuitive observation that a statistical analysis of aggregated data (combining multiple individual datasets) can lead to dramatically different conclusions from analyses of each dataset taken individually, i.e., that the whole appears to disagree with the parts. Simpson's paradox is easiest to grasp through an example. In the classic "Berkeley gender bias case" [4], the University of California at Berkeley was sued for gender bias against women applicants based on the aggregate 1973 admission figures (44% men admitted vs. 35% women)—an observational dataset. The much higher male figure appeared to be damning. However, when individually looking at the men *vs.* women, admission rate for each department, the rate was in fact similar for both sexes (and even in favor of women in most departments). The lower overall acceptance rate for women was not due to gender bias, but to the tendency of women to apply to more competitive departments, which have a lower admission rate in general. Thus, the association between gender and admission rate in the aggregate data could almost entirely be explained through strong association of these two variables with a third, confounding variable, the department. When controlling for the confounder, the association between the two first variables dramatically changes. This type of phenomenon is referred to as Simpson's paradox.

Because of the inherent heterogeneity of GO data, Simpson's paradox can manifest itself in GO analyses. This illustrates the importance of recognizing and controlling for potential biases and confounders.

1.2 The Inherent Incompleteness of the Gene Ontology (Open World Assumption)

The Gene Ontology is a representation of the current state of knowledge; thus, it is very dynamic. The ontology itself is constantly being improved to more accurately represent biology across all organisms. The ontology is augmented as new discoveries are made. At the same time, the creation of new annotations occurs at a rapid pace, aiming to keep up with published work. Despite these efforts, the information contained in the GO database, that is, the ontology and the association of ontology terms with genes and gene products, is necessarily incomplete. Thus, absence of evidence of function does not imply absence of function.[1] This is referred to as the Open World Assumption [5, 6].

Associations between genes/gene products and GO terms ("annotations") are made via various methods: some manual, some automated based on the presence of protein domains or because they belong to certain protein families [7]. Annotations can also be transferred to orthologs by manual processes [8], or automatically (e.g., [9, 10], reviewed in ref. 11). There are currently over 210 million annotations in the GO database. Despite these massive efforts to provide the widest possible coverage of gene products annotated, users should not expect each gene product to be annotated.

A further challenge is that the incompleteness in the GO is very uneven. Interestingly, the more comprehensively annotated parts of the GO can also pose challenges, presenting users with seemingly contradictory information (*see* Subheading 3.2).

The inherent incompleteness of GO creates problems in the evaluation of computational methods. For instance, overlooking the Open World Assumption can lead to inflated false positive rates in the assessment of gene function prediction tools [6]. However, there are ways of coping with this uncertainty. For instance, it is possible to gauge the effect of incomplete annotations on conclusions by thinning annotations [12], or analyzing successive, increasingly complete database releases [13, 14].

2 Gene Ontology Structure

One potential source of bias is that not all parts of the GO have the same level of details. This has a strong implication on measuring the similarity of GO annotations (Chap. 12 [2]). For instance, sister terms (terms directly attached to a common parent term) can be semantically very similar or very different in different parts of the GO structure, which has been called the "shallow annotation problem" (e.g., [15, 16]). This problem can partly be mitigated by the use of information-theoretic measures of similarity, instead of merely counting the number of edges separating terms, at the expense of requiring a considerable number of relevant annotations from which the frequency of co-occurrence of terms can be estimated (more details in Chap. 13 [3]).

[1] Proteins whose function is uncharacterized are annotated to the root of the ontology, which formally means "this protein is associated with *some* molecular function, biological process, or cellular component, but a more specific assertion cannot be made". This annotation is associated with the evidence code "No biological Data available" (ND). The absence of annotation indicates that no curator has reviewed the literature for this gene product.

2.1 Understanding Relationships Between Ontological Concepts

The GO is structured as a graph, and one pitfall of using the GO is to ignore this structure. Recall that each term is linked to other terms via different relationships (*see* Chaps. 1 [1] and 3 [17] for introductions to ontologies and GO annotations). These relationships need to be taken into account when using GO for data analysis.

Some relationships, such as "is a" and "part of", are *transitive*, which means that any protein annotated to a specific term is also implicitly annotated to all of its parents.[2] An illustration of this is a "serine/threonine protein kinase activity": it is a child of "protein kinase activity" with the relationship "is a". The transitivity of the relation means that the association between the protein and the term "serine/threonine protein kinase activity" and all its parents has the same meaning: the protein associated with "serine/threonine protein kinase activity" has this function, and it also has the more general function "protein kinase."

On the other hand, relations such as "regulates" are *non-transitive*. This implies that the semantics of the association of a gene to a GO term is not the same for its parent: if A is part of B, and B regulates C, we cannot make any inferences about the relationship between C and A. The same is true for positive and negative regulation. To illustrate, if we follow the term "peptidase inhibitor activity" (GO:0030414) to its parents, one of the terms encountered is "proteolysis" via a combination of "is a", "part of", and "regulates" relations. However, a "peptidase inhibitor activity" does not *mediate* proteolysis, but quite the contrary (Fig. 1). Thus, any logical reasoning on the ontology should take transitivity into account.

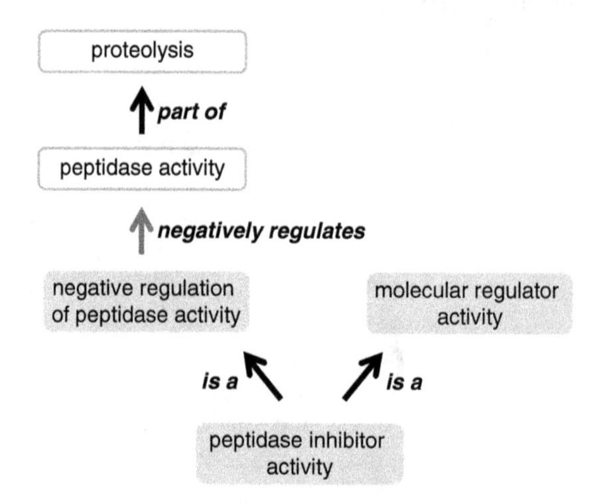

Fig. 1 Example of transitive (*black arrows*) and non-transitive (*red arrow*) relationships between classes. A protein annotated to "peptidase inhibitor activity" term does not imply it has a role in "proteolysis," since the link is broken by the non-transitive relation *negatively regulates*

[2] With the exception of "NOT" annotations, for which the transitivity applies to *children* terms, not *parents* (*see* also Subheading 3.2).

The relation "*has part*" is the inverse of "*part of*", and connects terms in the opposite direction. Because of this, it generates cycles in the ontology. The relation "*occurs in*" connects molecular function terms to the cellular components in which they occur. Thus, taking these relationships into account, it is possible to deduce additional cellular component annotations from molecular function annotations, without requiring additional experimental or computational evidence.

It important to know that there are three version of the GO ontology available: GO-basic, GO, and GO-plus.[3] Only the GO-basic file is completely acyclic. Therefore, applications requiring the traversal of the ontology graph usually assume that the graph is acyclic; hence, the GO-basic file should be used. The different GO ontology files are discussed in more detail in Chap. 11 [18].

2.2 Inter-ontology Links and Their Impact on GO Enrichment Analyses

The "part of" relation, when linking terms across the different *aspects* of the Gene Ontology (molecular function to biological process, or biological process to cellular component, for instance), triggers an annotation to the second term, using the same evidence code and the same reference, but "GOC" as the source of the annotation ("field 15 of the annotation file, *see* (Chap. 3 [17] for a description of the contents of the annotation file). For example, a DNA ligase activity annotation will automatically trigger an annotation to the biological process DNA ligation. The advantage of having these annotations inferred directly from the ontology is that it increases the annotation coverage by making annotations that may have been overlooked by the annotator when making the primary annotation. However, these inter-ontology links trigger a large number of annotations: there are currently 12 million annotations to 7 million proteins in the GO database. Changes in the structure of these links (as any change in the ontology), can potentially have a large impact on the annotation set. Indeed, Huntley et al. [19] reported that in November 2011, there was a decrease of ~2500 manually and automatically assigned annotations to the term "transcription, DNA-dependent" (GO:0006351) due to the removal of an inter-ontology link between this term and the Molecular Function term "sequence-specific DNA binding transcription factor activity" (GO:0003700). Figure 2 shows the strong and sudden variation in the number of annotations with term "ATPase activity" (GO:0016887).

Such large changes in GO annotations can affect GO enrichment analyses, which are sensitive to the choice of background distribution (Chap. 13 [3]; [20]). For instance, Clarke et al. [21] have shown that changes in annotations contribute significantly to changes in overrepresented terms in GO analysis. To mitigate this problem, researchers should analyze their datasets using the most

[3] http://geneontology.org/page/download-ontology

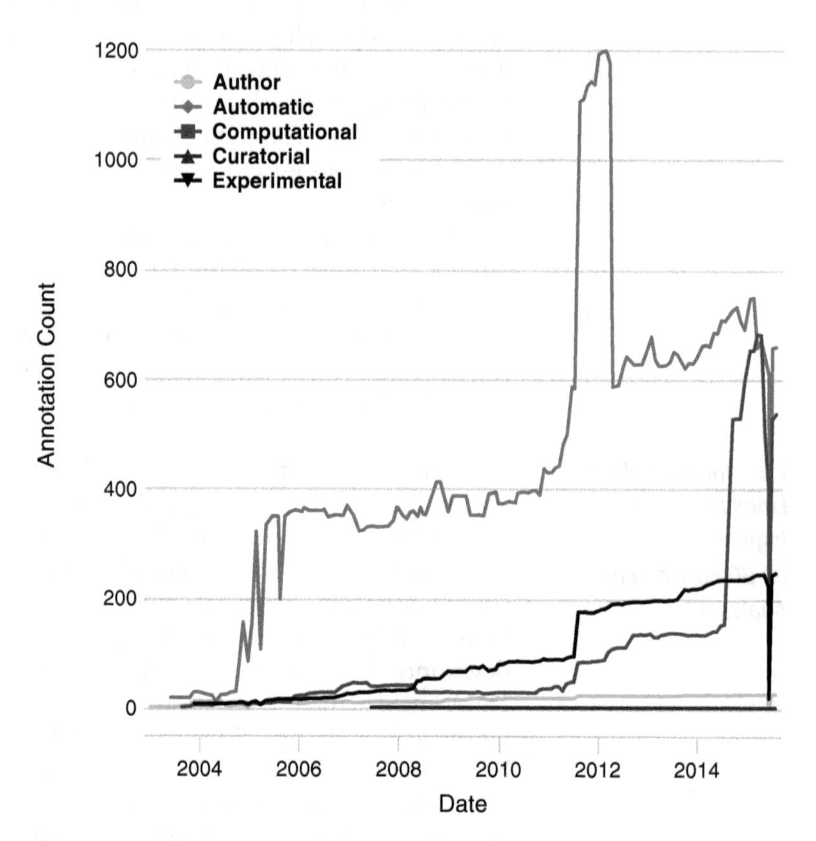

Fig. 2 Strong and sudden variation in the number of annotations with the GO term "ATPase activity" (GO:0016887) over time. Such changes can heavily affect the estimation of the background distribution in enrichment analyses. To minimize this problem, use an up-to-date version of the ontology/annotations and ensure that conclusions drawn hold across recent releases. Data and plot obtained from GOTrack (http://www.chibi.ubc.ca/gotrack)

up-to-date version of the ontology and annotations, and ensure that the conclusions they draw hold across multiple recent releases. At the time of the writing of this chapter, DAVID, a popular GO analysis tool, had not been updated since 2009 (http://david.abcc.ncifcrf.gov/forum/viewtopic.php?f=10&t=807). Enrichment analyses performed with it may thus identify terms whose distribution has substantially changed irrespective of the analysis of interest. The Gene Ontology Consortium now links to the PantherDB GO analysis service (http://amigo.geneontology.org/rte) [22]. This tool uses the most current version of the ontology and the annotations. Regardless of the tool used, researchers should disclose the ontology and annotation database releases used in their analyses.

3 Gene Ontology Annotations

Having discussed common pitfalls associated with the ontology structure, we now turn our attention to annotations. Understanding how annotations are done is essential to correctly interpreting the data. In particular, the information provided for each GO annotation extends beyond the mere association of a term with a protein (reference to Chap. 3 [17]). The full extent of this rich information, aimed to more precisely reflect the biology within the GO framework, is often overlooked.

3.1 Modification of Annotation Meaning by Qualifiers

The Gene Ontology uses three qualifiers that modify the meaning of association between a gene-product and a Gene Ontology term: These are "NOT", "contributes to", and "co-localizes with" (see documentation at http://geneontology.org/page/go-qualifiers).

The "contributes to" qualifier is used to capture the molecular function of complexes when the activity is distributed over several subunits. However, in some cases the usage of the qualifier is more permissive, and all subunits of a complex are annotated to the same molecular function even if they do not make a direct contribution to that activity. For example, the rat G2/mitotic-specific cyclin-B1 CCNB1 is annotated as contributing to histone kinase activity, based on data in [23], although it has only been shown to *regulate* the kinase activity of CDK1. Finding a cyclin annotated as having protein kinase activity may be unintuitive to users who fail to consider the "contributes to" qualifier.

The "co-localizes with" qualifier is used with two very different meanings: it first means that a protein is transiently or peripherally associated with an organelle or complex, while the second use is for cases where the resolution of an assay is not accurate enough to say that the gene product is a bona fide component member. Unfortunately, it is currently not possible to know which of the two meanings is meant in any given annotation.

3.2 Negative and Contradictory Results

The "NOT" qualifier is the one with the most impact, since it means that there is evidence that a gene product does *not* have a certain function. The "NOT" qualifier is mostly used when a specific function may be expected, but has shown to be missing, either based on closer review of the protein's primary sequence (e.g., loss of an active site residue) or because it cannot be experimentally detected using standard assays.

The existence of negative annotations can also lead to apparent contradictions. For instance, protein ARR2 in *Arabidopsis thaliana* is associated with "response to ethylene" (GO:0009723) both positively on the basis of a paper by Hass et al. [24] and negatively based on a paper by Mason et al. [25]. The latter discusses this contradiction as follows:

> Hass et al. [24] reported a reduction in the ethylene sensitivity of seedlings containing an arr2 loss-of-function mutation. By contrast, we observed no significant difference from the wild type in the seedling ethylene response when we tested three independent arr2 insertion mutants, including the same mutant examined by Hass et al. [24]. This difference in results could arise from differences in growth conditions, for, unlike Hass et al. [24], we used a medium containing Murashige and Skoog (MS) salts and inhibitors of ethylene biosynthesis.

Thus, in this case, the contradiction in the GO is a reflection of the primary literature. As Mason et al. note, this is not necessarily reflective of a mistake, as there can be differences in activity across space (tissue, subcellular localisation) and time (due to regulation), with some of these details not fully captured in the experiment or in its representation in the GO.

A NOT annotation may also be assigned to a protein that does not have an activity typical of its homologs, for instance the STRADA pseudokinase (UniProtKB:Q7RTN6); STRADA adopts a closed conformation typical of active protein kinases and binds substrates, promoting a conformational change in the substrate, which is then phosphorylated by a "true" protein kinase, STK11 [26]. In this case, the "NOT" annotation is created to alert the user to the fact that although the sequence suggests that the protein has a certain activity, experimental evidence shows otherwise.

In contrast to positive annotations, "NOT" annotations propagate to children in the ontology graph and not to parents. To illustrate, a protein associated with a negative annotation to "protein kinase activity" is not a tyrosine protein kinase either, a more specific term.

3.3 Annotation Extensions

As also described in Chap. 17 [27], the Gene Ontology has recently introduced a mechanism, the "annotation extensions", by which contextual information can be provided to increase the expressivity of the annotations [28]. Until recently, annotations had consisted of an association between a gene product and a term from one of the three ontologies comprising the GO. With this new knowledge representation model, additional information about the context of a GO term such as the target gene or the location of a molecular function may be provided.

Common uses are to provide data regarding the location of the activity/process in which a protein or gene product participates. For example, the role of Mouse opsin-4 (MGI:1353425) in rhodopsin mediated signaling pathway is biologically relevant in retinal ganglion cells. Annotation extensions also allow capture of dynamic subcellular localization, such as the *S. pombe* bir1 protein (SPCC962.02c), which localizes to the spindle specifically during the mitotic anaphase. The annotation extensions can also be used to capture substrates of enzymes, which used to be outside the scope of GO.

The annotation extension data is available in the AmiGO [29] and QuickGO [30] browsers, as well as in the annotation files compliant with the GAF2.0 format (http://geneontology.org/page/go-annotation-file-gaf-format-20). However, because annotation extensions are relatively new, guidelines are still being developed, and some uses are inconsistent across different databases. Furthermore, most tools have yet to take this information into account.

In effect, extensions of an annotation create a "virtual" GO class that can be composed of more than one "actual" GO class, and can be traced up through multiple parent lineages. Thus, just as with inter-ontology links, accounting for annotation extensions can result in a substantial inflation in the number of annotations, which needs to be appropriately accounted for in enrichment analyses and other statistical analyses that require precise specification of GO term background distribution.

3.4 Biases Associated with Particular Evidence Codes

Annotations are backed by different types of experiments or analyses categorized according to evidence codes (Chap. 3 [17]). Different types of experiments provide varying degrees of precision and confidence with respect to the conclusions that can be derived from them. For most experiment types, it is not possible to provide a quantitative measure of confidence. Evidence codes are informative but cannot directly be used to exclude low-confidence data.[4] Nonetheless, the different evidence codes are prone to specific biases.

Direct evidence. Taking these caveats into account, the evidence code inferred from direct assay (abbreviated as IDA in the annotation files) provides the most reliable evidence with respect to the how directly a protein has been implicated in a given function, as it names implies.

Mutant phenotype evidence. Mutants are extremely useful to implicate genes products in pathways and processes; however exactly how the gene product is implicated in the process/function annotated is difficult to assess using phenotypic data because such data are inherently derivative. Therefore, associations between gene products and GO terms based on mutant phenotypes (abbreviated as IMP in the annotation files) may be weak. The same caveat applies to annotations derived from mutations in *multiple* genes, indicated by evidence code "inferred from genetic interaction" (IGI).

Physical interactions. Evidence based on physical interactions (IPI; mostly protein–protein interactions) is comparable in confidence to a direct assay for protein binding annotations or for cellular components; however for molecular functions and biological

[4] An evidence confidence ontology has been proposed by Bastien et al. [31] but has yet to be adopted by the GO project.

processes, the evidence is of the type "guilt by association" and is of low confidence. Inferences based on expression patterns (IEP) are typically of low confidence. The presence of a protein in a specific subcellular localization, at a specific developmental stage, or associated with a protein or a protein complex can provide a hint to uncover a protein's role in the absence of other evidence, but without more direct evidence that information is very weak.

High-throughput experiments. Schnoes et al. [32] reported that annotations deriving from high-throughput experiments tend to consist of high-level GO terms, and tend to represent a limited number of functions. This artificially decreases the information content of these terms, since they are frequently annotated, and artificially decreased information content affects similarity analyses. This potentially has a large impact, since a significant fraction of the annotations in the GO database are derived from these types of analyses (as much as 25 %, according to Schnoes et al., who used the operational definition of a high throughput paper as one in which over 100 proteins were annotated). The GO does not currently record whether particular experimental annotations may be derived from high-throughput methods, but this may change in the future.

Biases from automatic annotation methods. The GO association file, containing the annotations, has information regarding the method used to assign electronic annotations. The annotations can be assigned by a large number of different methods. Examples include domain functions, as assigned for example by InterPro, by Enzyme Commission numbers being associated with an entry, by BLAST, by orthology assignment, etc. Note that this information is not provided as an evidence code, but as a "reference code". The list of methods and their associated reference code is available at http://www.geneontology.org/cgi-bin/references.cgi. The large number of electronic annotations can also make them have a disproportionate impact on the results. Most analysis tools allow for the inclusion or exclusion of electronic annotations, but not at the more fine-grained level of the particular method. It is nevertheless possible to use the combination of evidence code plus reference (available at: http://www.geneontology.org/cgi-bin/references.cgi) to automatically deepen the evidence type, see https://raw.githubusercontent.com/evidenceontology/evidenceontology/master/gaf-eco-mapping.txt).

Note that a gene or gene product can have multiple annotations to the same term but with different evidence. This can provide corroborating information on particular genes, but may also require appropriate normalization in statistical analyses of term frequency, as the frequency of terms that can be determined through multiple types of experiments may be artificially inflated. Furthermore, because different experiments can vary in their specificity—thus resulting in annotations at different levels of granularity for basically the same function—this redundancy only becomes conspicuous when the transitivity of the ontology structure is appropriately taken into account.

For more discussion on evidence codes, and their use in quality control pipelines, refer to Chap. 18 [33].

3.5 Differences Among Species

There can be substantial differences in the nature and extent of GO annotations across different species. For instance, zebrafish is heavily studied in terms of developmental biology and embryogenesis while the rat is the standard model for toxicology. These differences are reflected in the frequency of GO terms across species, which can vary considerably across species [34]. This has important implications on enrichment analyses and other statistical analyses requiring a background distribution of GO annotations. For instance, consider an experiment trying to establish the biological processes associated with a particular zebrafish protein by identifying its interaction partners and performing an enrichment analysis on them. If we naively use the entire database as background, the interaction partners might appear to be enriched in developmental genes simply because this class is over-represented in general in zebrafish. Instead, one should use zebrafish gene-related annotations only as background [20].

3.6 Authorship Bias

Other biases are less obvious but can nevertheless be strong and thus have a high potential to mislead. Recently, sets of annotations derived from the same scientific article were shown to be on average much more similar than annotations derived from different papers (Fig. 3; [34]). For instance, Nehrt et al. compared the

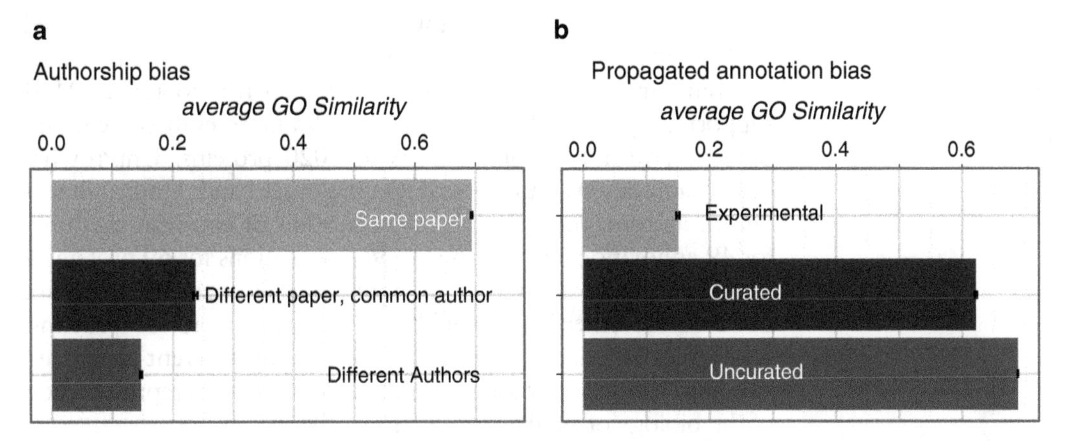

Fig. 3 (**a**) Average GO annotation similarity (using the measure of Schlicker et al. [35] between homologous genes, considering experimental annotations partitioned according to the provenance; (**b**) Average GO annotation similarity between homologous genes, partitioned according to their GO annotation evidence tags (Experimental: evidence code EXP and subcategories; Uncurated: evidence code IEA; Curated: all other evidence codes). Figure adapted from ref. 34

functional similarity of orthologs (genes related through speciation) across different species and paralogs (genes related through duplication) within the same species, and observed a much higher level of functional conservation among the latter [36]. However, this difference was almost entirely due to the fact that the GO functional annotations of same-species paralogs are ~50 times more likely to be derived from the same paper than orthologs; when controlling for authorship and other biases, the difference in functional similarity between same-species paralogs and orthologs vanished and even became in favor of orthologs [34].

Note that the difference is smaller but remains significant if we compare annotations established from different papers, but with at least one author in common, with annotations from different articles with no author in common.

3.7 Annotator Bias

Just as systematic differences among investigators can lead to the authorship bias, systematic differences in the way GO curators capture this information can lead to annotator bias. These annotator biases can in part be attributed to different annotation focus, but also to different interpretation or application of the GO annotation guidelines (http://geneontology.org/page/go-annotation-policies).

UniProt provides annotations for all species, which allows us to assess the effect of annotator (or database) bias. If we compare UniProt annotations for mouse proteins with those done by the Mouse Genome Informatics group (MGI), we see that comparable fractions of proteins are annotated using the different experimental evidence codes, with mutant phenotypes being the most widely used (78% of experimental annotations in MGI, versus 63% in UniProt), followed by direct assays (20% of annotations in MGI and 32% in UniProt).

However when we look at which GO terms are annotated based on phenotypes (IMP and IGI) by the two groups, we notice a large difference in the terms annotated. The top term annotated by MGI supported by the IMP evidence code is "in utero embryonic development", with 1170 annotations to 1020 proteins. UniProt has only 4 annotations for this term. On the other hand, UniProt has as one of its top-annotated classes "regulation of circadian rhythm", for 49 annotations to 38 proteins; 96 annotations for 69 proteins if we also include annotations to more specific, descendant terms. MGI on the other hand, only has 18 annotations for 19 proteins. This indicates that the annotations provided by different groups are biased towards specific aspects, and are not a uniform representation of the biology of all gene products in a species.

3.8 Propagation Bias

Another strong and perhaps surprising bias lies in the very different average GO similarity between electronic annotations compared with between experimental annotations. Indeed, if we consider homologous genes, their similarity in terms of electronic annotations tend to be much higher than in terms of experimental annotations, with curated annotations lying in-between ([34]; Fig. 3). A likely explanation for this phenomenon is that electronic annotations are typically obtained by inferring annotations among homologous sequences, a process that can only increase the average functional similarity of homologs.

Because of this homology inference bias, one must exercise caution when drawing conclusions from sets of genes whose annotations might have different proportions of experimental vs. electronic annotations. For instance, this would be the case when comparing annotations from model organisms with those from non-model organisms (the latter being likely to consist mostly of electronic annotations obtained through propagation).

More subtly, because function conservation is generally believed to correlate with sequence similarity, many computational methods preferentially infer function among phylogenetically close homologs. This bias can thus confound analyses attempting to gauge the conservation of gene function across different levels of species divergence.

3.9 Imbalance Between Positive and Negative Annotations

As discussed above, both our knowledge of gene function and its representation in the GO remain very incomplete. We have already discussed the pitfalls of ignoring this fact altogether (closed vs. open world assumption), or assuming similar term frequencies across species. But the extent of missing data varies along other dimensions as well: for example it can depend on how easy it is to experimentally establish a particular function and how interesting the potential function might be. The problem is particularly acute in the case of negative annotations, because they can be even more difficult to establish than their positive counterparts (e.g., a negative result can also be due to inadequate experimental conditions, differences in spatiotemporal regulation, etc.) *and* they are often perceived as being less useful, and certainly less publishable. As a result, currently less than 1 % of all experimental annotations are negative ones in UniProt-GOA [37]. This imbalance causes problems with training of machine learning algorithms [38]. Rider et al. [39] investigated the reliability of typical machine learning evaluation metrics (area under the “receiver operating characteristic” (ROC) curve, area under the precision-recall curve) under different levels of missing negative annotations and concluded that this bias could strongly affect the ranking obtained from the different metrics. Though this particular study adopted a closed world assumption, the effect of a varying proportion of negative annotations is likely to be even greater under the open world assumption.

4 Getting Help

This chapter provides a broad overview of some of the pitfalls associated with GO-based analysis. Table 1 summarizes the most important pitfalls users encounter using GO.

Users are advised to make use of a number of excellent resources provided by the GO consortium:

Table 1
Main pitfalls or biases discussed in the chapter and their remedies

Pitfall or bias	Remedy
Wrongly assume that absence of annotation implies absence of function.	Account for the fact that both ontology and annotations are necessary incomplete, for instance by assessing the impact of incompleteness on one's analyses and findings.
Not all directed edges in the ontology structure have the same meaning: depending on their type, the relationship they represent may or may not be transitive.	The transitivity of each type of relations must be taken into account when reasoning over the GO. "Is a" and "part of" are transitive, but "regulates" is not.
To yield meaningful results, GO enrichment analyses require accurate specification of the background distribution, which can vary substantially across releases, species, etc.	Specify the actual background distribution used in the analysis of interest. Short of this, ensure that the enrichment analysis is performed on consistent database release and subsets of species, terms, etc. To test the robustness of results, consider repeating the analysis using several releases of GO ontology/annotation databases. Avoid tools that are not regularly updated.
Inter-ontology links and annotation extensions can result in large variations in the number of annotations. Furthermore, annotation extensions may not be consistently implemented, if at all, across analyses tools or workflows.	Keep track of database releases in analyses. If they are relevant, make sure that annotation extensions are implemented consistently.
Qualifiers such as "NOT" or "co-localizes with" are important parts of a gene annotation in that they fundamentally change the meaning of annotations. Because only a small minority of all annotations have qualifiers, such errors can easily go unnoticed.	Remember to take into account qualifiers. When using tools or software libraries, make sure that these take qualifiers into account as well.
Annotations are supported by different types of evidence (categorized by evidence codes). The annotations associated with each code vary in their scope, specificity, and number. These differences can confound some analyses.	Take evidence code into account. In statistical analyses, consider the distribution of annotations in terms of evidence codes, and, if needed, control for this potential confounder.

(continued)

Table 1 (continued)

Pitfall or bias	Remedy
Different species tend to have very different types of annotations. For instance, model species have many more experiment-based annotations.	When performing statistical analyses or using information-theoretic similarity measures, use species-specific frequencies of GO term.
Experiment-based annotations derived from the same research article tend to be more similar than annotations derived from different articles. Similar trends hold for annotations derived from same versus different authors, and same versus different annotators.	Control for authorship bias in analyses that may have varying proportion of annotations stemming from the same article, lab, or annotation team.
Because annotations are preferentially propagated among closely related sequences, electronic annotations can confound analyses seeking to characterize relationships between evolution and function.	Restrict such analyses to experiment-based annotations. Avoid circularity.
There are many more positive annotations than negative annotations. As a result, standard accuracy measures used by machine learning methods may be misleading ("class imbalance problem").	Consider false-positive and false-negative rates separately. Focus on subset of data for which the class imbalance problem is less pronounced.

- The GO website http://geneontology.org
- The GO FAQ http://geneontology.org/faq-page
- The GO team are eager to help with your problems: e-mail go-help@geneontology.org
- The wider bioinformatics community can be consulted via sites like Biostars—see the GO tag https://www.biostars.org/t/go/
- The GO community can be contacted on Twitter at @news4go

5 Conclusion

This chapter surveys some of the main pitfalls and biases of the Gene Ontology. The number of potential issues, summarized in Table 1, may seem daunting. Indeed, as discussed at the start of this chapter, there are some inherent risks in working with observational data. However, simple remedies are available for many of these (Table 1). By understanding the subtleties of the GO, controlling for known confounders, trying to identify unknown ones, and cautiously proceeding forward, users can make the most of the formidable resource that is the GO.

Acknowledgements

We thank Natasha Glover, Rachel Huntley, Suzanna Lewis, Chris Mungall, and Paul Thomas for detailed and helpful feedback on the manuscript. PG acknowledges National Institutes of Health/ National Human Genome Research Institute grant HG002273. CD acknowledges Swiss National Science Foundation grant 150654 and UK BBSRC grant BB/M015009/1. Open Access charges were funded by the University College London Library, the Swiss Institute of Bioinformatics, the Agassiz Foundation, and the Foundation for the University of Lausanne.

References

1. Hastings J (2016) Primer on ontologies. In: Dessimoz C, Škunca N (eds) The gene ontology handbook. Methods in molecular biology, vol 1446. Humana Press. Chapter 1
2. Pesquita C (2016) Semantic similarity in the gene ontology. In: Dessimoz C, Škunca N (eds) The gene ontology handbook. Methods in molecular biology, vol 1446. Humana Press. Chapter 12
3. Bauer S (2016) Gene-category analysis. In: Dessimoz C, Škunca N (eds) The gene ontology handbook. Methods in molecular biology, vol 1446. Humana Press. Chapter 13
4. Bickel PJ, Hammel EA, O'connell JW (1975) Sex bias in graduate admissions: data from Berkeley. Science 187:398–404
5. Thomas PD, Wood V, Mungall CJ et al (2012) On the use of gene ontology annotations to assess functional similarity among orthologs and paralogs: a short report. PLoS Comput Biol 8:e1002386
6. Dessimoz C, Skunca N, Thomas PD (2013) CAFA and the Open World of protein function predictions. Trends Genet 29:609–610
7. Burge S, Kelly E, Lonsdale D et al (2012) Manual GO annotation of predictive protein signatures: the InterPro approach to GO curation. Database:bar068
8. Gaudet P, Livstone MS, Lewis SE et al (2011) Phylogenetic-based propagation of functional annotations within the Gene Ontology consortium. Brief Bioinform 12:449–462
9. Vilella AJ, Severin J, Ureta-Vidal A et al (2008) EnsemblCompara GeneTrees: complete, duplication-aware phylogenetic trees in vertebrates. Genome Res 19:327–335
10. Altenhoff AM, Škunca N, Glover N et al (2015) The OMA orthology database in 2015: function predictions, better plant support, synteny view and other improvements. Nucleic Acids Res 43:D240–D249
11. Rentzsch R, Orengo CA (2009) Protein function prediction--the power of multiplicity. Trends Biotechnol 27:210–219
12. Škunca N, Dessimoz C (2015) Phylogenetic profiling: how much input data is enough? PLoS One 10:e0114701
13. Škunca N, Altenhoff A, Dessimoz C (2012) Quality of computationally inferred gene ontology annotations. PLoS Comput Biol 8:e1002533
14. Jiang Y, Clark WT, Friedberg I et al (2014) The impact of incomplete knowledge on the evaluation of protein function prediction: a structured-output learning perspective. Bioinformatics 30:i609–i616
15. Sevilla JL, Segura V, Podhorski A et al (2005) Correlation between gene expression and GO semantic similarity. IEEE/ACM Trans Comput Biol Bioinform 2:330–338
16. Mistry M, Pavlidis P (2008) Gene Ontology term overlap as a measure of gene functional similarity. BMC Bioinformatics 9:327
17. Gaudet P, Škunca N, Hu JC, Dessimoz C (2016) Primer on the gene ontology. In: Dessimoz C, Škunca N (eds) The gene ontology handbook. Methods in molecular biology, vol 1446. Humana Press. Chapter 3
18. Munoz-Torres M, Carbon S (2016) Get GO! retrieving GO data using AmiGO, QuickGO, API, files, and tools. In: Dessimoz C, Škunca N (eds) The gene ontology handbook. Methods in molecular biology, vol 1446. Humana Press. Chapter 11
19. Huntley RP, Sawford T, Martin MJ et al (2014) Understanding how and why the Gene Ontology and its annotations evolve: the GO within UniProt. GigaScience 3:4
20. Rhee SY, Wood V, Dolinski K et al (2008) Use and misuse of the gene ontology annotations. Nat Rev Genet 9:509–515

21. Clarke EL, Loguercio S, Good BM et al (2013) A task-based approach for Gene Ontology evaluation. J Biomed Semantics 4(Suppl 1):S4
22. Mi H, Muruganujan A, Casagrande JT et al (2013) Large-scale gene function analysis with the PANTHER classification system. Nat Protoc 8:1551–1566
23. Granada JF, Ensenat D, Keswani AN et al (2005) Single perivascular delivery of mitomycin C stimulates p21 expression and inhibits neointima formation in rat arteries. Arterioscler Thromb Vasc Biol 25:2343–2348
24. Hass C, Lohrmann J, Albrecht V et al (2004) The response regulator 2 mediates ethylene signalling and hormone signal integration in Arabidopsis. EMBO J 23:3290–3302
25. Mason MG, Mathews DE, Argyros DA et al (2005) Multiple type-B response regulators mediate cytokinin signal transduction in Arabidopsis. Plant Cell 17:3007–3018
26. Baas AF, Boudeau J, Sapkota GP et al (2003) Activation of the tumour suppressor kinase LKB1 by the STE20-like pseudokinase STRAD. EMBO J 22:3062–3072
27. Huntley RP, Lovering RC (2016) Annotation extensions. In: Dessimoz C, Škunca N (eds) The gene ontology handbook. Methods in molecular biology, vol 1446. Humana Press. Chapter 17
28. Huntley RP, Harris MA, Alam-Faruque Y et al (2014) A method for increasing expressivity of Gene Ontology annotations using a compositional approach. BMC Bioinformatics 15:155
29. T. Gene and Ontology Consortium (2010) The Gene Ontology in 2010: extensions and refinements. Nucleic Acids Res 38:D331–D335
30. Binns D, Dimmer E, Huntley R et al (2009) QuickGO: a web-based tool for Gene Ontology searching. Bioinformatics 25:3045–3046
31. Bastian FB, Chibucos MC, Gaudet P et al (2015) The Confidence Information Ontology: a step towards a standard for asserting confidence in annotations. Database:bav043
32. Schnoes AM, Ream DC, Thorman AW et al (2013) Biases in the experimental annotations of protein function and their effect on our understanding of protein function space. PLoS Comput Biol 9:e1003063
33. Chibucos MC, Siegele DA, Hu JC, Giglio M (2016) The evidence and conclusion ontology (ECO): supporting GO annotations. In: Dessimoz C, Škunca N (eds) The gene ontology handbook. Methods in molecular biology, vol 1446. Humana Press. Chapter 18
34. Altenhoff AM, Studer RA, Robinson-Rechavi M et al (2012) Resolving the ortholog conjecture: orthologs tend to be weakly, but significantly, more similar in function than paralogs. PLoS Comput Biol 8:e1002514
35. Schlicker A, Domingues FS, Rahnenführer J et al (2006) A new measure for functional similarity of gene products based on Gene Ontology. BMC Bioinformatics 7:302
36. Nehrt NL, Clark WT, Radivojac P et al (2011) Testing the ortholog conjecture with comparative functional genomic data from mammals. PLoS Comput Biol 7:e1002073
37. Huntley RP, Sawford T, Mutowo-Meullenet P et al (2015) The GOA database: gene ontology annotation updates for 2015. Nucleic Acids Res 43:D1057–D1063
38. Kotsiantis S, Kanellopoulos D (2006) Handling imbalanced datasets: a review, Annual Symposium on Foundations of Computer Science
39. Rider AK, Johnson RA, Davis DA et al (2013) Classifier evaluation with missing negative class labels. In: Advances in Intelligent Data Analysis XII. Springer, Berlin, pp 380–391

13

Semantic Similarity in the Gene Ontology

Catia Pesquita

Abstract

Gene Ontology-based semantic similarity (SS) allows the comparison of GO terms or entities annotated with GO terms, by leveraging on the ontology structure and properties and on annotation corpora. In the last decade the number and diversity of SS measures based on GO has grown considerably, and their application ranges from functional coherence evaluation, protein interaction prediction, and disease gene prioritization.

Understanding how SS measures work, what issues can affect their performance and how they compare to each other in different evaluation settings is crucial to gain a comprehensive view of this area and choose the most appropriate approaches for a given application.

In this chapter, we provide a guide to understanding and selecting SS measures for biomedical researchers. We present a straightforward categorization of SS measures and describe the main strategies they employ. We discuss the intrinsic and external issues that affect their performance, and how these can be addressed. We summarize comparative assessment studies, highlighting the top measures in different settings, and compare different implementation strategies and their use. Finally, we discuss some of the extant challenges and opportunities, namely the increased semantic complexity of GO and the need for fast and efficient computation, pointing the way towards the future generation of SS measures.

Key words Gene ontology, Semantic similarity, Functional similarity, Protein similarity

1 Introduction

The graph structure of the Gene Ontology (GO) allows the comparison of GO terms and GO-annotated gene products by semantic similarity. Assessing similarity is crucial to expanding knowledge, because it allows us to categorize objects into kinds. Similar objects tend to behave similarly, which supports inference, a crucial task to support many applications including identifying protein–protein interactions [1], suggesting candidate genes involved in diseases [2] and evaluating the functional coherence of gene sets [3, 4].

Semantic similarity (SS) assesses the likeness in meaning of two concepts. It has been a subject of interest to Artificial Intelligence, Cognitive Science, and Psychology for the last few decades, and an important tool for Natural Language Processing. It has been used in this context to perform word sense disambiguation, determining discourse structure, text summarization and annotation, information extraction and retrieval, automatic indexing, lexical selection, and automatic correction of word errors in text [5].

Sometimes, research literature uses SS, relatedness, and distance as interchangeable terms, but they are in fact not identical. Semantic relatedness makes use of various relations between two concepts (i.e., hyponymic, hypernymic, meronymic, antonymic, and any kind of functional relations including has-part, is-made-of, and is-an-attribute-of). SS is more limited since it usually only makes use of hierarchical relations, such as hyponymy/hyperonymy (i.e., is-a), and synonymy. Most authors support that semantic distance is the opposite of similarity, but it is sometimes also used as the opposite of semantic relatedness.

The basis for much of the earlier research in SS is the WordNet, a large lexical database of the English language, freely available online. However, the last decade has witnessed an explosion in the number of applications of SS to biomedical ontologies, and specifically in the GO [6]. The GO structure provides meaningful links between GO terms, based on the various relationships it establishes. This structure allows us to capture the similarity between GO terms. In general, the closer two terms are in the GO graph, the more similar their meaning is. Moreover, we can also determine the similarity between two GO-annotated gene products by expanding on this notion to compare sets of GO terms. This provides a measure of the functional similarity between two proteins, which has numerous applications in biomedical research.

The remainder of this chapter provides an overview of SS between GO terms and gene products annotated with GO terms, the different kinds of approaches used in this research area, the issues that affect their performance and evaluation and challenges and future directions.

2 SS Measures

A SS measure can be defined as a function that, given two ontology terms or two sets of terms annotating two entities, returns a numerical value reflecting the closeness in meaning between them [7]. For a theoretical framework for SS measures please refer to [8], where the core elements shared by most SS measures are identified and a foundation for the comparison, selection, and development of novel measures is laid out.

In the context of GO, SS measures can be applied to compute the similarity between two GO terms, *term similarity*, or to compute the similarity between two gene products each annotated with a set of GO terms, *gene product similarity*.

In recent years there have been several categorizations of SS measures [7, 9], and we advise readers to refer to both surveys for a more detailed classification and survey of SS measures and their applications.

2.1 Term Similarity

When considering SS between concepts organized in a taxonomy, as is the case of GO, there are two basic approaches: internal methods based on ontology structure and external methods based on external corpora.

The simplest structural methods calculate distance between two nodes as the number of edges in the path between them [10]. If there are multiple paths, the shortest path or an average of all possible paths can be used. For instance, in Fig. 1, the distance between *heme binding* and *anion binding* is 5. This measure depends only on the structure of the graph and it assumes that all semantic links have the same weight. Accordingly, SS is defined as the inverse score of the semantic distance. This edge-counting approach is intuitive and simple but disregards the depth of the nodes, since it considers

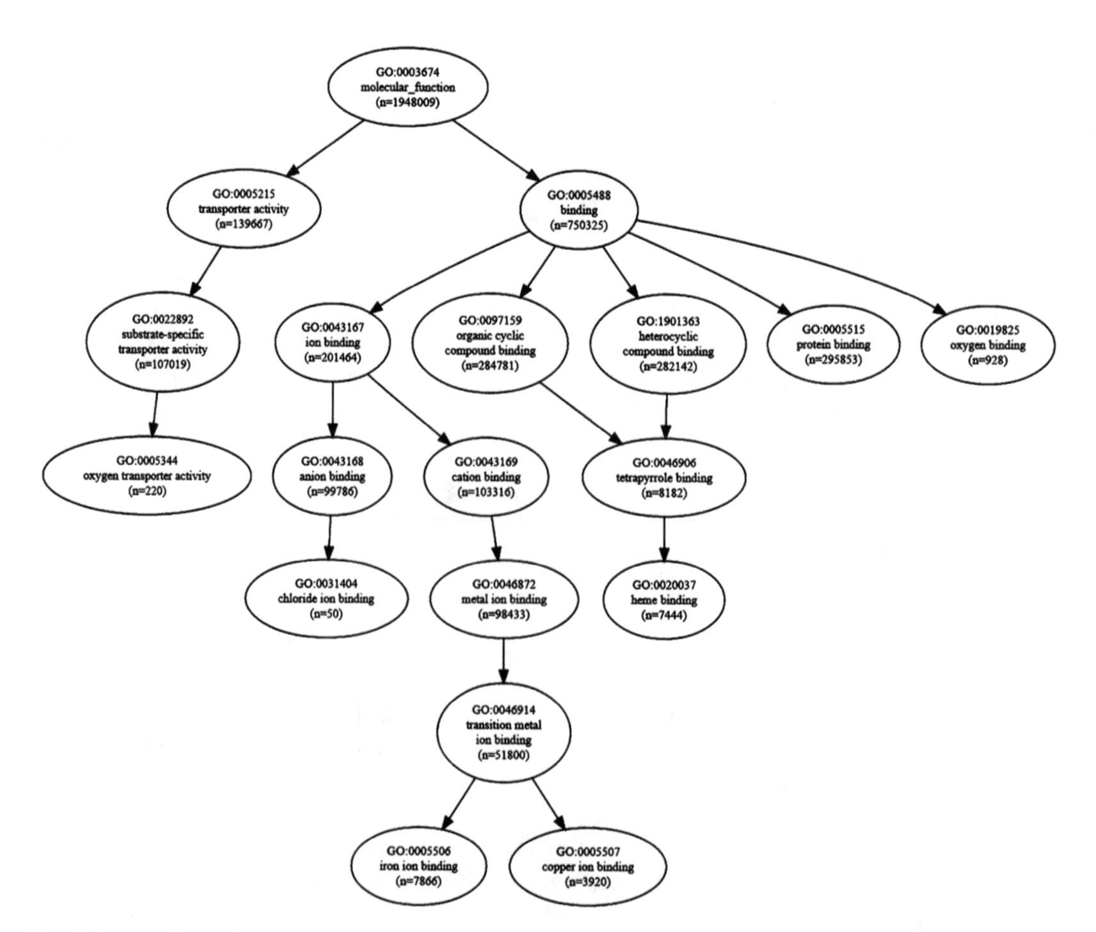

Fig. 1 Subgraph of GO covering the annotations of hemoglobin subunit alpha and hemocyanin II proteins. The number of gene products annotated to each term in GOA (January, 2016) are indicated by *n*

paths of equal length to equate to the same degree of similarity, regardless if they occur near the root or deeper in the ontology. For instance, in Fig. 1, the classes *transport* and *binding* are at a distance of two edges, the same distance that separates *iron ion binding* and *copper ion binding*.

To overcome this limitation of equal distance edges, some approaches give edges different weights to reflect some degree of hierarchical depth. It is intuitive that the deeper the level in the taxonomy, the smaller the conceptual distance, so weights are reduced according to depth. Other factors can be used to determine weights for edges such as node density and type of link.

However these methods have two important limitations, they rely heavily on the assumption that nodes and edges in an ontology are uniformly distributed and that nodes at the same level correspond to the same semantic distance, which are untrue in the case of GO. For instance, in Fig. 1, although *oxygen binding* and *ion binding* are both at a depth of 2, the former is a more specific concept and is actually a leaf node. More recent approaches attempt at mitigating some of these issues using for instance the depth of the lowest common ancestor (LCA) [11], distance to nearest leaf node [12], and depth of distinct GO subgraphs [1]. Related approaches, also based on the structure of the ontology, combine distance metrics with node structural properties, such as number of subclasses and distance to the lowest common ancestor between the terms [13].

External methods typically make use of information-theoretic principles. This type of approach has been demonstrated to be less sensitive or not at all to the issue of link density variability [14], i.e., that the ontology graph may be unbalanced and edges linking nodes may not be evenly distributed, so that the same depth or distance indicate a different level of specificity or similarity. Information content (IC)-based measures are based on the intuition that the similarity between two concepts can be given by the extent to which they share information.

The IC of a concept c is a measure of how likely the concept is to occur, which can be quantified as the negative log likelihood, $-log\ p(c)$ where $p(c)$ is the probability of occurrence of c in a specific corpus, usually estimated by the annotation frequency in the Gene Ontology Annotation database. A normalized version of IC was introduced in [15], whereby IC values are expressed in a range of uniformly scaled values, making them easier to interpret. Taking Fig. 1 again as an example, the frequency of annotation of *binding* is 750,325/1,948,009, making its IC 1.38 and its normalized IC 0.066.

When the concept of IC is applied to the common ancestors two terms have, it can be used to quantify the information they share and thus measure their SS. There are two main approaches for doing this: the most informative common ancestor (MICA technique), in which only the common ancestor with the highest

IC is considered [14]; and the disjoint common ancestors (DCA technique), in which all disjoint common ancestors (the common ancestors that do not subsume any other common ancestor) are considered. There are several methods to compute the DCA [16–18], which allow IC-based measures to take into account multiple common ancestors.

Several measures have been used to measure the information shared by two GO terms. The simplest of these measures, Resnik's, takes the IC of the MICA as the similarity between two terms, and was among the first to be applied to GO [19]. The MICA of *chloride ion binding* and *iron ion binding* is *ion binding*, making the Resnik similarity between these terms to be 0.066. Other measures combine the IC of terms with the IC of the MICA and weight them according to the MICA's IC [20].

More recently, hybrid measures that combine both edge and IC-based strategies have been proposed [21]. Corpus-independent IC measures have also been proposed, based on number of descendants [22], depth and descendants [23] and on the notion of entropy [24].

2.2 Gene Product Similarity

Since gene products can be annotated with several GO terms within each of the three GO categories, gene product SS measures need to compare sets of terms rather than single terms. Several approaches have been proposed for this, most following one of two strategies: pairwise or groupwise.

Pairwise approaches take the individual similarities between all terms annotating two gene products and combine them into a global measure of functional similarity. Any term similarity measure can be applied with this strategy, where each gene product is represented by its set of direct annotations. Typical combination strategies include the average, maximum, or sum, and these can be applied to every pairwise combination of terms from the two sets or only the best-matching pair for each term.

Groupwise approaches calculate gene product similarity directly by one of three approaches: set, graph, or vector. Set approaches consider only direct annotations and are calculated using set similarity techniques. Set-based measures are limited in that they do not take into account the shared ancestry between GO terms. Graph approaches represent gene products as the subgraphs of GO corresponding to all their annotations. Functional similarity is then calculated either using graph-matching techniques or by less computationally intensive approaches such as set similarity. This approach takes into account all annotations (direct and inherited) providing a more comprehensive model of the annotations. Vector approaches represent gene products in vector space, with each term corresponding to a dimension, and functional similarity is calculated using vector similarity measures. Groupwise approaches can also make use of the IC of terms, by using it to weigh set similarity computations, such as simGIC [15], which compares two sets of terms based on a IC-weighted Jaccard similarity; as scalar values in vectors, such as IntelliGO [25], which combines IC and the evidence content of annotations; or to compute the IC of shared subgraphs, such as the SS measure proposed in [14].

3 Issues and Challenges in SS

Guzzi et al. [9] have identified several issues affecting SS measures, which they categorize into external issues, which are usually related to annotation corpora, and internal issues, inherent to the design of the measures. They do however recognize that both kinds of issues can be entangled, for instance when measures make erroneous assumptions about the corpora.

The most relevant external issues are the shallow annotation problem, the annotation length bias, and the use of Evidence Codes. The shallow annotation problem stems from the fact that many proteins are only annotated to very general GO terms, thus for instance two proteins can share 100% of their terms and still be very dissimilar. SS measures need to account for this issue, which can be especially relevant in the electronic annotations. Nevertheless, the quality and specificity of these annotations has been increasing over the years [26].

The annotation length bias refers to the positive correlation between SS scores and the number of annotations that some measures produce. This is due to the fact that annotations are not uniformly distributed among the proteins within an annotation corpus (and also vary among different organisms corpora), with some proteins being very well annotated while others have a single annotation. Both of these issues stem from incomplete annotations, which have been shown to have a significant impact in the performance of information-theoretic measures [27]. Finally, SS approaches need to be aware of the impact that using electronic annotations (evidence code IEA) can have.[1] Although in general the use of IEA annotations has a positive or null effect on the measures performance, in some cases and particularly when employing the maximum combination approach over pairwise similarities it can have a detrimental effect and decrease the measure's ability to capture similarity as conveyed by evaluation metrics [9, 17].

There are three levels at which internal issues can occur: term specificity, term similarity, and gene product similarity. At the term specificity level, both typically used approaches (term depth and IC) have their advantages and drawbacks. IC-based measures can be affected by the corpus bias effect [29] whereby rarely used but generic terms possess a high IC but are not biologically specific.

[1] Please *see* Chap. 3 [28] for more information on evidence codes.

This issue is particularly relevant when using specific corpora that may be incomplete. Term depth measures on the other hand, while being independent of annotation corpora, are unable to handle the fact that terms at the same depth rarely have the same biological specificity, given the fact that GO's regions have varying node and edge density.

At the term similarity level, distance-based measures suffer from the same issues as term depth term specificity. Moreover, since most measures rely on the concept of common ancestors to measure similarity between two terms, SS measures need to define the set of common ancestors over which similarity is computed. While the most informative common ancestor (or lowest common ancestor in the case of edge-based measures) is commonly used and usually provides good performance, it has been argued that measures taking into account all ancestors or a selection of them can more adequately portray the whole gamut of function.

At the gene product similarity level, and in particular for pairwise measures, special care needs to be taken when choosing a combination approach. The maximum approach is unsuitable to assess their global similarity, since it focuses on the single most similar aspect. The average approach, on the other hand, by making an all-against-all comparison of the terms of two gene products, produces counterintuitive results for gene products with multiple distinct functional aspects. For instance, two gene products both annotated with the same two unrelated terms, *t1* and *t2*, will be 50% similar under the average approach, because similarity will be calculated between both the matching (*t1–t1*,*t2–t2*) and the opposite (*t1–t2*,*t2–t1*) terms of the two gene products. The best-match approach would rely on comparing just (*t1–t1*,*t2–t2*), since these are the best-matching term pairs in the annotations set. The best-match average approach generally provides a better performance by considering all terms but only the most significant matches.

4 Evaluating and Comparing SS Measures

Evaluating the reliability of SS measures or determining the best measure for each application scenario is still an open question since there is no gold standard. Furthermore, each of the existing measures formalizes the notion of function similarity in slightly different ways and for that reason it is not possible to define what the best SS measure would be, since it becomes a subjective decision. Ultimately, SS measures attempt to capture functional similarity based on GO annotations, so one possible solution is to compare SS measures to other measures or proxies of functional similarity. These include sequence similarity, family similarity, protein–protein interactions, functional modules and complexes, and expression profile similarity. Table 1 details the best performing measures

Table 1
Best performing SS measures according to different protein similarity measures or proxies. Sequence, Pfam, and ECC similarity correspond to correlation evaluated using CESSM

Similarity proxy or measure	Best performing SS measures
Sequence similarity	SSDD [13], SimGIC [15], HRSS [21]
Pfam similarity	SORA [23], SSDD, SimGIC
ECC similarity	SSDD, HRSS, SORA
Expression similarity	TCSS [1], SimGIC, SimIC, Best-Match-Avg (Resnik [15])
Protein–protein interaction	TCSS, SimIC, Max(Resnik)

Results compiled from refs. 9, 13, 21, 23, 30

for each aspect according to a recent survey of literature. Although more classic measures of SS such as Resnik still provide top results in some settings, it is the newer generation of measures that provides the best results. And if until recently [9] GOA-based IC measures were regarded as the best performing measures for most settings, the new wave of more complex structural-based measures, such as SSDD [13], SORA [23] and TCSS [1] are now on the lead, though closely followed by SimGIC. SSDD is based on the concept of semantic "totipotency" whereby terms are assigned values according to their distance to the root and the number of descendants for each of the levels in that path, and then similarity corresponds to the smallest sum of "totipotencies" along a path between two terms. SORA uses an IC based on structural information that considers depth and number of descendants, and then applies set similarity to gene products. TCSS divides the GO graph into subgraphs and considers gene products more similar if they belong to the same subgraph. We postulate that the recent success of structural and hybrid measures, is not only due to their ability to more accurately capture the complexity of the GO graph, but also due to the evolution of GO itself, which has grown considerably since the "classic" measures were proposed. Linear correlation to sequence similarity is one of the most used measures, and in general a positive correlation between sequence and SS has been found, particularly on binned data. Nonlinear regression analysis found that the normal cumulative distribution fits data for many different SS measures, confirming the positive yet, nonlinear agreement between sequence and SS [15]. Linear correlation has also been used to compare SS to Pfam-based and Enzyme Commission Class similarity.

One of the most relevant efforts in this area is the Collaborative Evaluation of Semantic Similarity Measures (CESSM) tool [30], which was created in 2009 to answer this need. It enables the comparison of new GO-based SS measures against previously published ones considering their relation to sequence, Pfam, and Enzyme Commission Class (ECC) similarity. Since its inception, CESSM has been adopted by the community and used to evaluate several novel SS measures.

The predictive power of SS measures in identifying protein–protein interactions is also commonly employed in SS evaluation [9]. In general SS measures are good predictors of PPI, but the most effective are groupwise or maximum combination approach measures. This is unsurprising given that proteins can interact when sharing a single functional aspect.

5 Tools

There are two main kinds of available tools to compute SS measures in GO: webservers, which typically provide easy to use solutions with fewer parametrizations possible; and software packages, which are more customizable, though more complex to use.

Many of the recently proposed SS measures provide specific webservers, but some online tools provide a wider array of measures, such as ProteInOn [31], FunSimMat [32], or GOssToWeb [33]. These tools rely on their own GO and GOA versions, and though they can output similarity scores with an input of just GO terms or Uniprot accession numbers, these scores are based on the tool's ontology and annotation versions.

If a user needs more control over the parametrization of the input data, then the best option is to employ a software package. Options include R packages (e.g., GoSemSim [34]) or standalone programs (GOssTo [33]), which give the user more freedom in terms of ontology and annotation versions as well as in programmatic access or the computation of SS for larger datasets. A Java library has been recently developed for ontology-based SS calculations [35], which includes over 50 different SS measures and accepts input ontologies in a number of formats, including OWL, OBO, and RDF. This library is well suited for large input datasets, being able to run over 100 million comparisons in under 1 h. In the case of webtools, we advise readers to check their update frequency to ensure that recent versions of GO and the annotations are in use.

6 Challenges and Future Directions

The last decade has witnessed a growing interest in GO-based SS, with dozens of new measures being proposed and applied in different settings. Although measures have become increasingly sophisticated, there remain several challenges and opportunities.

GO-based SS measures are inherently dependent on GO's development and its use in annotations. Measures should evolve with GO, striving to provide ever more accurate metrics for gene product functional similarity. In recent years there have been several developments of GO which SS measures are still not exploring. For instance, the different kinds of regulatory and occurrence relationships, the categorization of evidence codes, logical definitions and internal and external cross-products, can all in principle be explored by SS approaches.

The need to provide more semantically sound measures of SS for biomedical ontologies has been argued [36], and though GO is commonly viewed as a DAG for a controlled vocabulary it is actually well axiomatized in OWL [37]. The presence of these axioms should be considered by SS measures, and the exploration of disjointness in SS has been recently proposed in ChEBI [38].

In general, the computational complexity of SS measures has not been addressed. Current GO-based SS applications happen in an offline context where computational speed is not a relevant factor. However, for applications such as similarity-based search, which so far are based on precomputed similarities [32], performance should be taken into consideration. In addition, the growth in size of biomedical datasets spurred by genomic scale studies in the last few years, also places further computational constraints on SS measures. The challenge of handling very large datasets is increasingly recognized, and recent implementations of SS measures allow for parallel computation [35], but the development of SS measures is not taking this issue into consideration *a priori*.

The next generation of SS measures should take into account these two aspects, on one hand, the possibility for increased complexity in SS measures to provide more accurate similarity scores, and on the other the need for efficient SS computation, and strive to achieve a balance between increased accuracy and efficiency.

7 Exercises

Consider the subgraph of GO represented in Fig. 1 and the number of annotations for each GO term it shows.

1. Calculate the IC of the term "heme binding" considering that the total universe of annotations corresponds to the number of annotations to the root term.
2. Transform the IC value calculated in 1 to a uniform scale [0,1]. Consider that the maximum IC is given to a term with a single annotated gene product, and an IC of zero corresponds to the IC of the root term, "molecular function."
3. Calculate the SS between the terms "chloride ion binding" and "iron ion binding," and "oxygen transporter activity," and

"tetrapyrrole binding," following the minimum edge distance measure.

4. Calculate Resnik's SS between the same terms as in c.
5. Calculate the similarity between the protein *hemoglobin subunit alpha* annotated with [ion iron binding, copper ion binding, protein binding, heme binding, oxygen binding, oxygen transporter activity], and the protein *hemocyanin II* annotated with [chloride ion binding, copper ion binding, oxygen transporter activity]:
   (a) Using the average of all pairwise Resnik's similarities
   (b) Using the maximum of all pairwise Resnik's similarities
   (c) Using the simGIC measure, which corresponds to the ratio between sum of the IC of the shared terms between the two proteins and the sum of the IC of the union of all terms between the two proteins.
   (d) Compare the obtained results with your perception of the actual functional similarity between the two proteins.

References

1. Jain S, Bader GD (2010) An improved method for scoring protein-protein interactions using semantic similarity within the gene ontology. BMC Bioinformatics 11(1):562
2. Li X, Wang Q, Zheng Y, Lv S, Ning S, Sun J, Li Y (2011) Prioritizing human cancer microRNAs based on genes' functional consistency between microRNA and cancer. Nucleic Acids Res 39(22):e153
3. Richards AJ, Muller B, Shotwell M, Cowart LA, Rohrer B, Lu X (2010) Assessing the functional coherence of gene sets with metrics based on the Gene Ontology graph. Bioinformatics 26(12):i79–i87
4. Bastos HP, Clarke LA, Couto FM (2013) Annotation extension through protein family annotation coherence metrics. Front Genet 4:201
5. Budanitsky A, Hirst G (2001) Semantic distance in WordNet: an experimental, application-oriented evaluation of five measures. In Workshop on WordNet and other lexical resources, vol 2, pp 2–2
6. Ashburner M, Ball CA, Blake JA, Botstein D, Butler H, Cherry JM, Sherlock G et al (2000) Gene ontology: tool for the unification of biology. Nat Genet 25(1):25–29
7. Pesquita C, Faria D, Falcao AO, Lord P, Couto FM (2009) Semantic similarity in biomedical ontologies. PLoS Comput Biol 5(7):e1000443
8. Harispe S, Sánchez D, Ranwez S, Janaqi S, Montmain J (2014) A framework for unifying ontology-based semantic similarity measures: a study in the biomedical domain. J Biomed Inform 48:38–53
9. Guzzi PH, Mina M, Guerra C, Cannataro M (2012) Semantic similarity analysis of protein data: assessment with biological features and issues. Brief Bioinform 13(5):569–585
10. Rada R, Mili H, Bicknell E, Blettner M (1989) Development and application of a metric on semantic nets. IEEE Trans Syst Man Cybernet 19(1):17–30
11. Yu H, Gao L, Tu K, Guo Z (2005) Broadly predicting specific gene functions with expression similarity and taxonomy similarity. Gene 352:75–81
12. Cheng J, Cline M, Martin J, Finkelstein D, Awad T, Kulp D, Siani-Rose MA (2004) A knowledge-based clustering algorithm driven by gene ontology. J Biopharm Stat 14(3):687–700
13. Xu Y, Guo M, Shi W, Liu X, Wang C (2013) A novel insight into Gene Ontology semantic similarity. Genomics 101(6):368–375
14. Resnik P (1999) Semantic similarity in a taxonomy: an information-based measure and its application to problems of ambiguity in natural language. J Artif Intell Res (JAIR) 11:95–130

15. Pesquita C, Faria D, Bastos H, Ferreira AE, Falcão AO, Couto FM (2008) Metrics for GO based protein semantic similarity: a systematic evaluation. BMC Bioinformatics 9(Suppl 5):S4
16. Couto FM, Silva MJ, Coutinho PM (2005) Semantic similarity over the gene ontology: Family correlation and selecting disjunctive ancestors. Proceedings of the ACM conference in information and knowledge management
17. Couto FM, Silva MJ (2011) Disjunctive shared information between ontology concepts: application to Gene Ontology. J Biomed Semantics 2:5
18. Zhang SB, Lai JH (2015) Semantic similarity measurement between gene ontology terms based on exclusively inherited shared information. Gene 558(1):108–117
19. Lord P, Stevens R, Brass A, Goble C (2003) Investigating semantic similarity measures across the Gene Ontology: the relationship between sequence and annotation. Bioinformatics 19:1275–1283
20. Schlicker A, Domingues FS, Rahnenführer J, Lengauer T (2006) A new measure for functional similarity of gene products based on gene ontology. BMC Bioinformatics 7:302
21. Wu X, Pang E, Lin K, Pei ZM (2013) Improving the measurement of semantic similarity between gene ontology terms and gene products: insights from an edge-and IC-based hybrid method. PLoS One 8(5):e66745
22. Seco N, Veale T, Hayes J (2004) An intrinsic information content metric for semantic similarity in wordnet. ECAI, pp 1089–1090
23. Teng Z, Guo M, Liu X, Dai Q, Wang C, Xuan P (2013) Measuring gene functional similarity based on group-wise comparison of GO terms. Bioinformatics:btt160
24. Warren A, Setubal J (2012) Using entropy estimates for DAG-based ontologies. In Proceedings of the 15th bio-ontologies special interest group meeting of ISMB 2012
25. Benabderrahmane S, Smail-Tabbone M, Poch O, Napoli A, Devignes MD (2010) IntelliGO: a new vector-based semantic similarity measure including annotation origin. BMC Bioinformatics 11(1):588
26. Škunca N, Altenhoff A, Dessimoz C (2012) Quality of computationally inferred gene ontology annotations. PLoS Comput Biol 8(5):e1002533
27. Jiang Y, Clark WT, Friedberg I, Radivojac P (2014) The impact of incomplete knowledge on the evaluation of protein function prediction: a structured-output learning perspective. Bioinformatics 30(17):i609–i616
28. Gaudet P, Škunca N, Hu JC, Dessimoz C (2016) Primer on the gene ontology. In: Dessimoz C, Škunca N (eds) The gene ontology handbook. Methods in molecular biology, vol 1446. Humana Press. Chapter 3
29. Mistry M, Pavlidis P (2008) Gene ontology term overlap as a measure of gene functional similarity. BMC Bioinformatics 9:327
30. Pesquita C, Pessoa D, Faria D, Couto F (2009) CESSM: collaborative evaluation of semantic similarity measures. In: JB2009: challenges in bioinformatics, vol 157, p 190
31. Faria D, Pesquita C, Couto FM, Falcão A (2007) Proteinon: a web tool for protein semantic similarity. Department of Informatics, University of Lisbon
32. Schlicker A, Albrecht M (2008) FunSimMat: a comprehensive functional similarity database. Nucleic Acids Res 36(Suppl 1):D434–D439
33. Caniza H, Romero AE, Heron S, Yang H, Devoto A, Frasca M et al (2014) GOssTo: a stand-alone application and a web tool for calculating semantic similarities on the Gene Ontology. Bioinformatics 30(15):2235–2236
34. Yu G, Li F, Qin Y, Bo X, Wu Y, Wang S (2010) GOSemSim: an R package for measuring semantic similarity among GO terms and gene products. Bioinformatics 26(7):976–978
35. Harispe S, Ranwez S, Janaqi S, Montmain J (2014) The semantic measures library and toolkit: fast computation of semantic similarity and relatedness using biomedical ontologies. Bioinformatics 30(5):740–742
36. Couto FM, Pinto HS (2013) The next generation of similarity measures that fully explore the semantics in biomedical ontologies. J Bioinforma Comput Biol 11(05): 1371001
37. Mungall CJ, Dietze H, Osumi-Sutherland D (2014) Use of OWL within the Gene Ontology. Proceedings of the 11th international workshop on OWL: experiences and directions. Riva del Garda, Italy, 2014
38. Ferreira JD, Hastings J, Couto FM (2013) Exploiting disjointness axioms to improve semantic similarity measures. Bioinformatics 29(21):2781–2787

14

Visualizing GO Annotations

Fran Supek and Nives Škunca

Abstract

Contemporary techniques in biology produce readouts for large numbers of genes simultaneously, the typical example being differential gene expression measurements. Moreover, those genes are often richly annotated using GO terms that describe gene function and that can be used to summarize the results of the genome-scale experiments. However, making sense of such GO enrichment analyses may be challenging. For instance, overrepresented GO functions in a set of differentially expressed genes are typically output as a flat list, a format not adequate to capture the complexities of the hierarchical structure of the GO annotation labels.

In this chapter, we survey various methods to visualize large, difficult-to-interpret lists of GO terms. We catalog their availability—Web-based or standalone, the main principles they employ in summarizing large lists of GO terms, and the visualization styles they support. These brief commentaries on each software are intended as a helpful inventory, rather than comprehensive descriptions of the underlying algorithms. Instead, we show examples of their use and suggest that the choice of an appropriate visualization tool may be crucial to the utility of GO in biological discovery.

Key words Gene Ontology, Visualization, Interpretation, Redundancy, Enrichment, Tools

1 Introduction

We have entered the era of massive data sets in biology. A variety of experimental and computational techniques can produce readouts for many genes—or whole genomes—simultaneously. Moreover, we can also assign rich functional annotations to most of the genes of interest. Such a wealth of data is accompanied with challenges in interpretation.

In this chapter, we focus on methods that visualize long lists of Gene Ontology (GO) terms [1]. The methods we survey take as input a flat list of GO terms, often accompanied by some user-supplied measure of statistical significance or importance. Visualization methods summarize such lists to distil the most relevant information. Finally, these methods produce various styles of visualization that can aid interpretation.

First, we examine the challenges related to understanding large lists of GO terms; second, we provide a systematic overview of the published methods that address these challenges; third, we discuss different visualization styles these methods use; and fourth, we give usage examples for a selection of these tools.

2 Understanding Large Lists of Genes and Their Gene Ontology Labels

A classical example of a large biological dataset are gene expression measurements by RNA-Seq, which monitor the genome-wide changes in transcriptional regulation between experimental conditions. Typically, tens or hundreds of genes will be upregulated or downregulated in response to a particular treatment. This indicates that a systems-level change in the experimental model has occurred, which may be described by examining the common properties of the genes whose expression was altered. Do these genes participate in the same metabolic or signaling pathways? Do they perform similar biochemical functions? Do their protein products co-localize in the cell? Formally, such sets of genes are subjected to statistical tests for enrichment for various functional categories [2]. The gene functions tested are typically described by Gene Ontology (GO) terms [3], although alternatives such as KEGG Pathways or CORUM protein complexes can be used.

Of note, such GO enrichment analyses are by no means restricted to experiments measuring changes in gene expression, nor to experimental data in general. Any list of genes for which interpretation is sought can be described using enriched GO terms and it could, for instance, derive from comparative genomics. In particular, one could perform an evolutionary analysis to look at biological roles of gene families that have expanded in a certain eukaryotic lineage, e.g., [4]. Similarly, a researcher may wish to describe the overall functional repertoire in a newly sequenced genome, while comparing to existing genomes of related organisms.

2.1 Challenges in Interpreting Lists of Enriched GO Terms

As Chap. 3 [5] describes, the GO is a hierarchical structure, wherein the individual terms can have not only multiple descendants, but also multiple parents; more formally, GO is a directed graph; the basic version of the GO is also a directed *acyclic* graph (Chap. 3[1] [5]; Fig. 1). This complex structure, along with its large size—the GO has thousands of nodes—make it challenging to display the part(s) of the GO of interest. For instance, a list of GO terms found to be enriched in a gene expression experiment could be concentrated in one part of the GO graph.

A further complication is that such lists of interesting GO terms tend to be large, meaning that many different biological processes or molecular functions may appear to be affected in the experiment.

[1] http://geneontology.org/page/download-ontology

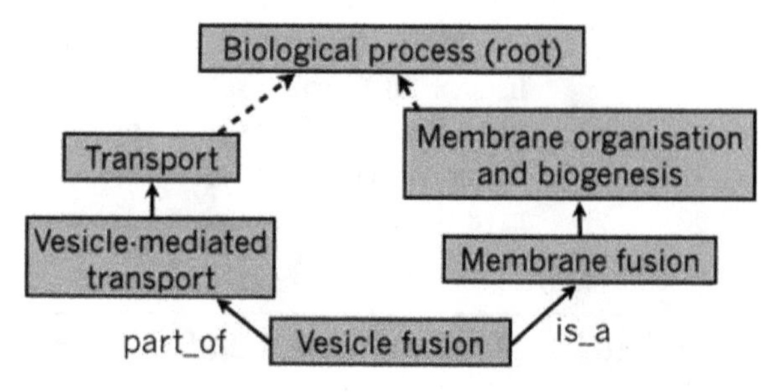

Fig. 1 A subset of the Gene Ontology Directed Acyclic Graph (DAG) for the GO term "vesicle fusion" (GO:0006906). The GO is a DAG: terms are nodes, while the relations are edges. Two main relation types between terms are "is_a" and "part_of." More specific terms are found deeper in the graph. Thus, if a gene product is annotated with a GO term, it is by definition also annotated with all the parent terms of that GO term

One reason for this is that the GO itself is designed and developed to describe nuances in gene function as exhaustively as possible; consequently, many of the GO terms will be partially redundant. For instance, many of the genes participating in "translation" (GO:0006412) are also structurally a part of "ribosome" (GO:0005840).

In addition to the inherent redundancy of the GO, responses of biological systems to experimental perturbation often genuinely involve coordinated activity of many related and/or overlapping subsystems. For example, replicating cells facing DNA damage may upregulate "nucleotide-excision repair" (GO:0006289) to help fix the lesions, but at the same time resorting to "error-prone translesion synthesis" (GO:0042276) to ensure DNA replication finishes.

2.2 Visualizing the GO to Facilitate Insight and Avoid Biases

GO term enrichment analyses often result in lists of significant GO terms that are both long and redundant, hampering interpretation. Various methods to visualize such lists may help investigators spot dominant trends in the data, leading to novel biological insight. Such visualizations mostly operate by different ways of grouping and displaying similar GO terms together, wherein the structure of the GO defines what is similar and what is not (see *semantic similarity analysis* below). In its simplest form, this involves displaying a part of the GO hierarchy with the GO terms of interest highlighted and their parent–child relationships shown. Displaying also the user-supplied experimental data may help prioritize which GO terms, among many similar ones, are of higher interest.

We suggest that having an unbiased way to algorithmically organize GO terms derived from experimental data helps prevent unintentional biases in interpretation. If unaware of the overall semantic structure in the set of significant GO terms, the investigator may pick one or two GO terms in the list that "make sense," in terms of fitting with their expectations. By visualizing the interrelationships between the GO terms alongside the statistical support for each in the experimental data could help avoid focusing on outlying—and perhaps spurious—results. In addition, one could be made aware of the common pitfall where one GO term is chosen, while other similarly statistically supported terms are ignored. Finally, and very importantly, a good visualization is also an effective means of presenting summaries of scientific results, whether in papers, presentations or posters.

3 Overview of the GO Visualization-Related Tools

Here we systematize and describe the currently available tools for visualizing sets of GO annotations. Additionally, we highlight three of these tools in more detail. The tools and the underlying methods they implement can be classified thusly:

1. *Interactive GO browsers.* Tools for interactively browsing the entire GO and also the genes known to be annotated with chosen GO terms. Importantly, these do not take into account a user-supplied set of annotations of interest, e.g., derived from an enrichment analysis of experimental data. Visualization is typically not emphasized and not configurable. See AmiGO [6] and QuickGO [7]. Of note, OLSVis [8] can display other biomedical ontologies in addition to the GO.
2. *Network visualization tools.* These are not particular to the GO, but can display any kind of graph, including the GO or a part thereof. The visualization options are highly configurable; however, since these tools were not designed specifically for GO, they tend to be more complicated to use. See Cytoscape [9], Gephi [10], and Pajek [11].
   (a) Of note, there are Cytoscape plugins specialized for handling groups of GO terms: EnrichmentMap [12] and BINGO [13].
3. *GO visual overlays.* Tools that can visualize an interesting subset of the GO, and display some additional data about each shown GO term. Typically, this involves coloring the GO terms by the enrichments or p-values determined from user-supplied gene lists (these tools tend to also perform the GO enrichment analysis). They display the terms arranged by parent–child relationships, in a tree-like visual layout. Examples include GOrilla [14], GRYFUN [15], GOFFA [16], and SimCT [17].
   (a) In addition to the GO, similar tools are available which can highlight the individual members in displayed KEGG pathways [18]; the pathways can also be shown in a KEGG BRITE functional hierarchy with FuncTree [19].
4. *Semantic similarity analysis.* Tools that examine the semantic similarity (redundancy) between various GO terms, including

those that are not linked by direct parent–child relationships. The similarities are used to organize a set of interesting GO terms into clusters and/or graphs, while simultaneously allowing highly redundant terms to be filtered out. The user can supply enrichments or p-values to prioritize results. Implemented in REVIGO [20] and RedundancyMiner [21].

(a) Some provisions for this are made in g:Profiler [22], which collapses similar GO terms.

(b) The Ontologizer [23] can perform a statistical test for enrichment that accounts for the parent–child redundancy [24] prior to visualizing results.

5. *Emerging methods.* These may involve display of the trends underlying a group of GO terms in a so-called "tag cloud" (with text in various colors and sizes), or in a tree map (a hierarchical organization of colored tiles), as in REVIGO [20] or GOSummaries [25]. Additionally, several tools now support the display of multiple GO enrichment analyses side-by-side; see BACA [26] or GOSummaries. SimCT [17] can display subtrees of other biomedical ontologies in addition to the GO.

4 Case Studies with Selected Tools

GOrilla [14] is a Web-based tool that can take two types of input: either a ranked list of genes or two lists, one with the target genes and the other with the background genes. As output, GOrilla produces a visualization that indicates which terms are significantly enriched.

We focus here on the enrichment analysis that takes a ranked list of genes. Briefly, the null hypothesis is that the occurrences of a GO term at various points in the ranked list are equiprobable. Lower *p*-values indicate a higher confidence for a GO term to be enriched towards the top of the list.

As an example analysis, we downloaded a dataset of transcription profiling by microarray of human peripheral blood mononuclear cells after a treatment with *Staphylococcus aureus* and incubation for different lengths of time [27], obtained from the Gene Expression Atlas [28] at http://www.ebi.ac.uk/gxa/experiments/E-GEOD-16837. In the GOrilla Web interface (http://cbl-gorilla.cs.technion.ac.il/), we set the p-value threshold to 10^{-3}, and the remaining settings were the defaults in the tool.

The display is shown in Fig. 2. Based on the color of the boxes, the user can visualize which GO terms are enriched, and the connecting lines describe their relationship to other terms in the GO graph.

REVIGO [20] analyzes large lists of significant GO terms and removes the redundant terms, in order to further narrow the search to a set of nonredundant and highly significant GO terms. Briefly, REVIGO creates clusters of GO terms that are semantically similar, and selects one representative for each cluster.

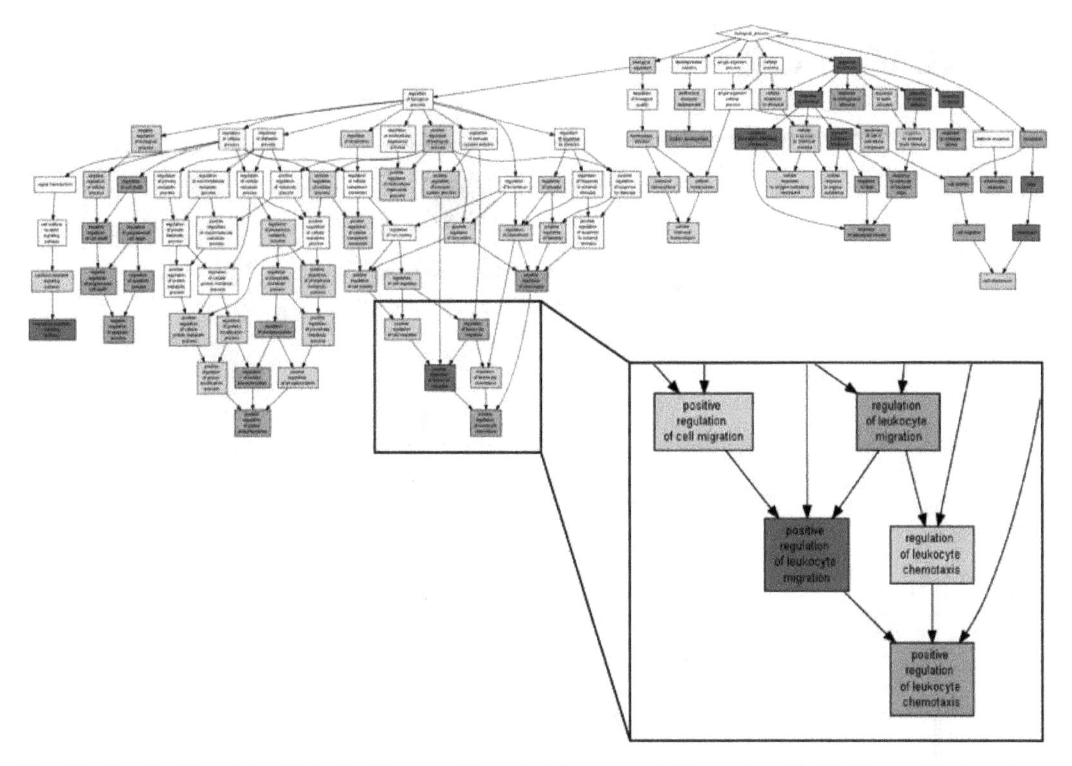

Fig. 2 A visualization of the Biological Process Gene Ontology annotations using GOrilla. The dataset used is a microarray transcription profiling of human peripheral blood mononuclear cells after treatment with *Staphylococcus aureus* (Expression Atlas dataset ID E-GEOD-16837). The GOrilla settings were left at default values: *p*-value threshold of $p < 10^{-3}$, organism *Homo sapiens* and running mode "single ranked list"

One possible input for REVIGO is a list of GO terms with the associated p-values, such as the output list from GOrilla. Alternatively, REVIGO can take as input any other list of GO terms, with or without associated numerical values, and provide various styles of visualization. First, a scatterplot that distributes the GO terms, represented as bubbles, in a 2D space that will put two GO terms closer together if they are more semantically similar. Second, an interactive graph that connects the user-supplied set of GO terms based on the structure of the GO hierarchy. Third, a TreeMap where terms are clustered and clusters displayed as colored tiles. Fourth, REVIGO provides a word cloud that highlights the most frequent keywords in the names and descriptions of the GO terms.

To perform the analysis, we used the setting in GOrilla to automatically forward its GO term enrichment results as a query to the REVIGO tool. In REVIGO, we used the default settings.

Fig. 3 (continued) while its size reflects the generality of the GO term in the UniProt-GOA database. (**b**) The table view shows the list of all the input GO terms: those shown in the scatterplot are written in regular font, while those labeled as redundant by REVIGO are shown in *gray italics*

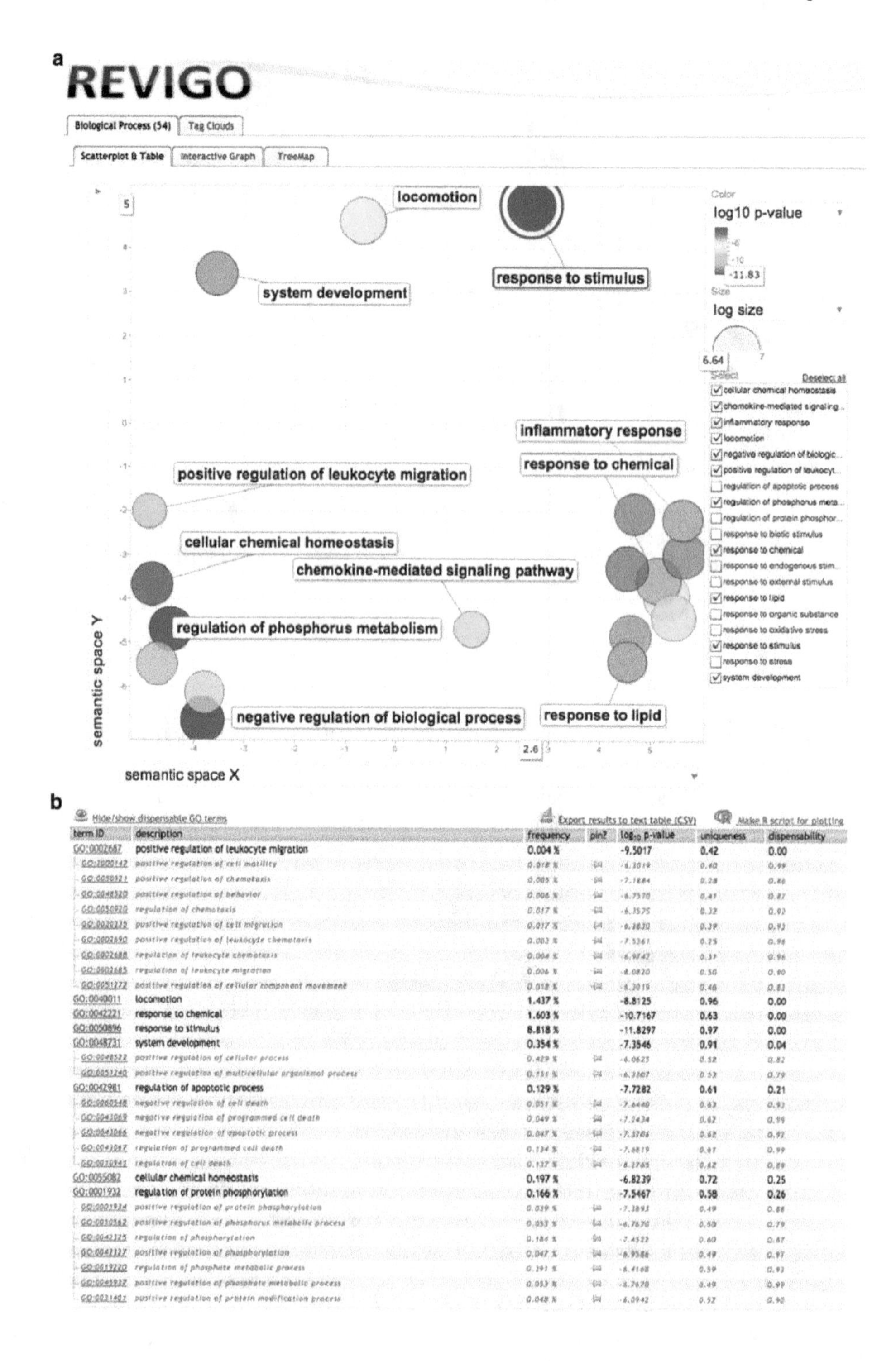

Fig. 3 Visualizations of Biological Process GO annotations using REVIGO: scatterplot and table views. The dataset used was imported from GOrilla (see legend of Fig. 2). We used the default settings of the REVIGO tool. (**a**) The scatterplot view visualizes the GO terms in a "semantic space" where the more similar terms are positioned closer together [20]. The color of the bubble reflects the p-value obtained in the GOrilla analysis,

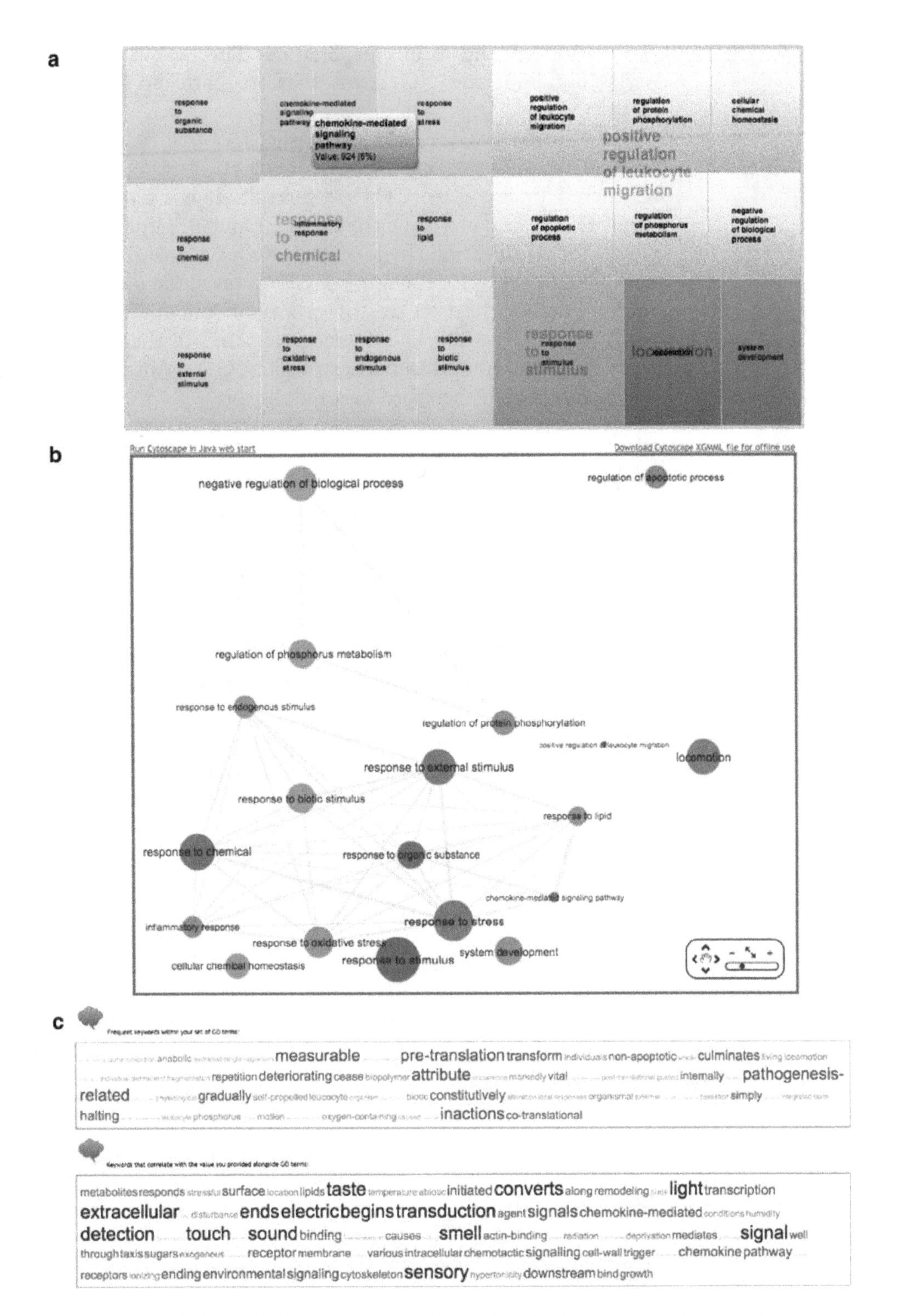

Fig. 4 Visualizations of Biological Process GO annotations using REVIGO: TreeMap (**a**), interactive graph (**b**) and word cloud views (**c**). The dataset used was imported from GOrilla (see legend of Fig. 2). We used the default settings of the tool

The results are shown in Figs. 3 and 4. The various visualization styles highlight the GO terms that are enriched in the input dataset.

RedundancyMiner [21] is another tool that focuses on non-redundant terms in a large list of enriched GO terms, producing a Clustered Image Map (CIM) as a result. It is a part of a larger pipeline: RedundancyMiner relies on GOminer input and on CIM miner for visualization. In particular, RedundancyMiner performs Fisher's exact tests for each pair of GO terms in the datasets, calculating whether the two sets of genes annotated with these GO terms are overlapping. A symmetrical matrix of these p-values is subsequently analyzed to arrive to a set of GO terms that are most independent, and therefore least redundant.

To perform the analysis, we started with the same file as for the two tools described above. First, we generated two files using a custom Python script: (1) a file containing all the genes in the array and (2) a file containing the genes that are over or underexpressed, labeled with "1" or "−1," respectively. Of note, Python is not necessary for RedundancyMiner and these files could be generated otherwise. Second, we put these files as input for the GOminer tool (http://discover.nci.nih.gov/gominer/GoCommandWebInterface.jsp). We selected the databases that contain *Homo sapiens* data and as the organism we set *H. sapiens*. The remaining parameters were the defaults in the tool. Third, we used the resulting folder as the working folder for RedundancyMiner and we ran the analysis in default mode. Finally, we visualized the resulting CIM file using cimMiner [29], available at http://discover.nci.nih.gov/cimminer/home.do, in single matrix mode.

Results of our example analysis are shown in Fig. 5. Even with the stringent threshold of requiring the $\log_2$ fold change greater than 5, similar trends in significant GO terms are visible as shown with the remaining two tools.

5 Choice of Visualization

Above, we have outlined some of the currently available software tools that can visualize a set of GO terms. We have also argued that a good visualization is an effective means of discovering underlying trends in the data in an unbiased fashion; an appropriate visual display is also imperative when communicating the results to others. The question of which software tool to apply should be addressed keeping these goals in mind. A related yet distinct question is which specific visualization method to choose. Here, we give a summary of the available options. Of note, the authors of this text are also the developers of REVIGO [20], a versatile visualization tool, which implements several of the approaches listed below.

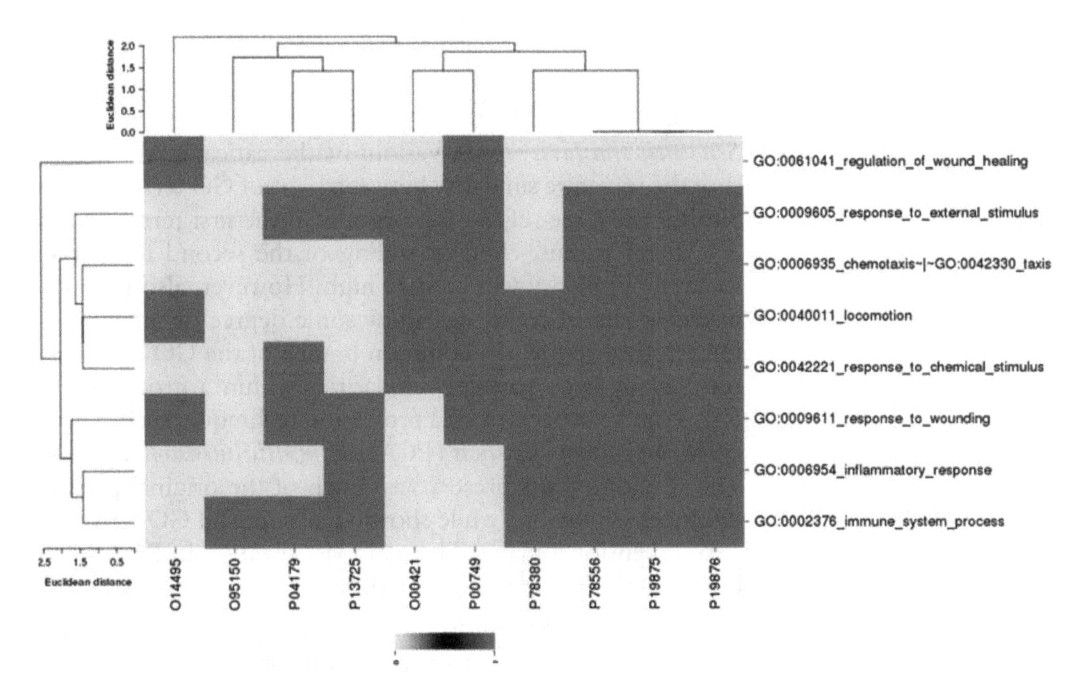

Fig. 5 A visualization of a set of Biological Process GO annotations using RedundancyMiner. The dataset used is a microarray transcription profiling of human peripheral blood mononuclear cells after treatment with *Staphylococcus aureus*. For this visualization, we focus on genes that had log2 fold change greater than 5

1. *Graphs/networks*. The GO graph consists of *nodes* (here, Gene Ontology terms) and *edges* (here, parent–child relationships), which connect the nodes and which have directionality. Nodes and edges can have multiple attributes that can be visualized. For instance, the enrichment of a GO term in a user's experiment may be shown as a color of a node (Fig. 2). Importantly, the spatial arrangement of the nodes on the final plot is called a *layout*, and is often created to suggest related clusters of nodes by placing similar nodes closer together. Such approaches are reviewed and demonstrated by Merico et al. [30]; tools like Cytoscape [9] support a variety of visual layouts.
   (a) A special case of a layout is a tree-like display that highlights the 'levels' in the Gene Ontology and the parent–child relationships between terms (e.g., Fig. 2). These levels (determining the *depth* of a node in the graph) are often used as a measure for how general the GO term is. However, this may be misleading in some instances—for example, the Molecular Function ontology is more shallow than the Biological Process ontology—and we therefore recommend the use of the *information content* (IC) measure [31] for this purpose. This is defined as the negative

logarithm of the relative frequency of the respective term annotations in some underlying database, such as the UniProt-GOA [32].

2. *Semantic similarity space.* Various mathematical methods measure the semantic similarity between pairs of GO terms, such as SimRel [33]; see ref. 34 for a review. If the first term in a pair is a direct parent, child or sibling of the second term, their semantic similarity will be very high. However, also the more distantly related terms will show some degree of similarity, as long as they reside in a common branch of the GO tree structure. Many such pairwise similarities within a group of GO terms can be processed by a projection technique, such as *principal components analysis* (PCA) or *multidimensional scaling.* The resulting plots preserve as much of the original pairwise distances as possible, while showing all supplied GO terms in a two-dimensional plane. The main visualization in REVIGO is based on this approach (Fig. 3a).
3. *Treemaps.* Hierarchical diagrams consisting of tiles subdivided into smaller tiles. Treemaps are good for interactive exploration, as they can be 'zoomed in' by clicking a tile and revealing finer levels of subdivisions. Here, tiles can be GO terms and the subdivisions their child terms. The tile sizes may correspond to some measure of importance of GO terms to the user, such as enrichment or p-values. REVIGO has an implementation of this visualization approach (Fig. 4b).
4. *Word clouds.* A display with text shown in various sizes and possibly colors. Here, the individual words or short phrases may be the names of the GO terms or some keywords associated to the GO terms. The text size/color may convey the importance to the user (enrichment), or in some instances generality of a GO term (see *information content* above). This visualization method is implemented in GOSummaries and REVIGO (Fig. 4c).
5. *Clustered Heatmaps.* Two-dimensional grids of values, wherein the rows and/or columns are clustered to reveal the 'block structure' in the data. Clustered heatmaps are often used for showing high-dimensional data in biology, but rarely so for GO terms. In fact, this could be done to show the GO terms' similarity based on what genes are annotated to them, or on the terms' semantic similarity (which is defined by the structure of the GO graph). An example implementation can be found in RedundancyMiner (Fig. 5).

In addition to the above, many of the tools specializing in GO enrichment testing (or in other analyses of large-scale biological data) often come bundled with visualizations that include GO as an important context. Examples include the Bioconductor packages GOexpress, GOfunction and GOSim. In addition, it is often possible to customize such displays in more detail by manually passing the GO data to a dedicated visualization software, such as the *ggplot2* package [35] in R, or to *gnuplot* software. For example, a specialized software to draw treemaps can be made to display GO enrichments from a biological experiment via a script that prepares the data in a correct format [36]. REVIGO will draw bubble charts where the GO terms are displayed in a semantic similarity space [20], and it can export a *ggplot2* script which is further customizable for e.g., font sizes, colors, and line styles; it can similarly export a graph to be further customized in Cytoscape.

6 Concluding Remarks and Outlook

In summary, we outline several tools that biologists can use to visualize sets of Gene Ontology terms and uncover novel and interesting trends in their experimental data. We anticipate that the future will bring even more massive biological data sets, which will have several consequences. First, the lists of interesting GO terms will grow in length, as larger sample sizes afford more statistical power to detect associations. Therefore, refinements of the existing approaches that address redundant GO terms [20, 21] will come in useful. Second, the visualization software will need to deal with more than a single list of enriched GO terms. While some current tools can display such results from multiple experiments side-by-side, e.g. BACA [26], tools will be needed that can integrate such lists and extract patterns across them. Finally, while GO is a prominent example of an ontology used by biologists, it is far from the only one [37]—over 100 biomedical ontologies exist that describe environments, phenotypes, and chemical entities (*see* Chap. 19) [38]. We foresee substantial developments in the tools that can summarize and visualize results of various biological experiments in the context of such emerging ontologies.

Acknowledgements

We acknowledge the support of the European Commission via grants MAESTRA (ICT-2013-612944) and InnoMol (FP7-REGPOT-2012-2013-1-316289), and of the Croatian Science Foundation grant MultiCaST (# 5660). Open Access charges were funded by the University College London Library, the Swiss Institute of Bioinformatics, the Agassiz Foundation, and the Foundation for the University of Lausanne.

References

1. Ashburner M, Ball CA, Blake JA et al (2000) Gene ontology: tool for the unification of biology. The Gene Ontology Consortium. Nat Genet 25:25–29
2. Rivals I, Personnaz L, Taing L et al (2007) Enrichment or depletion of a GO category within a class of genes: which test? Bioinformatics 23:401–407
3. du Plessis L, Skunca N, Dessimoz C (2011) The what, where, how and why of gene ontology--a primer for bioinformaticians. Brief Bioinform 12:723–735
4. Lespinet O, Wolf YI, Koonin EV et al (2002) The role of lineage-specific gene family expansion in the evolution of eukaryotes. Genome Res 12:1048–1059
5. Gaudet P, Škunca N, Hu JC, Dessimoz C (2016) Primer on the gene ontology. In: Dessimoz C, Škunca N (eds) The gene ontology handbook. Methods in molecular biology, vol 1446. Humana Press. Chapter 3
6. Carbon S, Ireland A, Mungall CJ et al (2009) AmiGO: online access to ontology and annotation data. Bioinformatics 25:288–289
7. Binns D, Dimmer E, Huntley R et al (2009) QuickGO: a web-based tool for Gene Ontology searching. Bioinformatics 25:3045–3046
8. Vercruysse S, Venkatesan A, Kuiper M (2012) OLSVis: an animated, interactive visual browser for bio-ontologies. BMC Bioinformatics 13:116
9. Smoot ME, Ono K, Ruscheinski J et al (2011) Cytoscape 2.8: new features for data integration and network visualization. Bioinformatics 27:431–432
10. Bastian M, Heymann S, Jacomy M (2009) Gephi: an open source software for exploring and manipulating networks. In: Third international AAAI conference on weblogs and social media
11. Batagelj V (2011) Exploratory social network analysis with Pajek (Structural analysis in the social sciences). Cambridge University Press, Cambridge
12. Merico D, Isserlin R, Bader GD (2011) Visualizing gene-set enrichment results using the Cytoscape plug-in enrichment map. Methods Mol Biol 781:257–277
13. Maere S, Heymans K, Kuiper M (2005) BiNGO: a Cytoscape plugin to assess overrepresentation of gene ontology categories in biological networks. Bioinformatics 21:3448–3449
14. Eden E, Navon R, Steinfeld I et al (2009) GOrilla: a tool for discovery and visualization of enriched GO terms in ranked gene lists. BMC Bioinformatics 10:48
15. Bastos HP, Sousa L, Clarke LA et al (2015) GRYFUN: a web application for GO term annotation visualization and analysis in protein sets. PLoS One 10:e0119631
16. Sun H, Fang H, Chen T et al (2006) GOFFA: gene ontology for functional analysis--a FDA gene ontology tool for analysis of genomic and proteomic data. BMC Bioinformatics 7(Suppl 2):S23
17. Herrmann C, Bérard S, Tichit L (2009) SimCT: a generic tool to visualize ontology-based relationships for biological objects. Bioinformatics 25:3197–3198
18. KEGG Mapper. http://www.genome.jp/kegg/tool/map_pathway2.html
19. Uchiyama T, Irie M, Mori H et al (2015) FuncTree: functional analysis and visualization for large-scale omics data. PLoS One 10:e0126967
20. Supek F, Bošnjak M, Škunca N et al (2011) REVIGO summarizes and visualizes long lists of gene ontology terms. PLoS One 6:e21800
21. Zeeberg BR, Liu H, Kahn AB et al (2011) RedundancyMiner: de-replication of redundant GO categories in microarray and proteomics analysis. BMC Bioinformatics 12:52
22. Reimand J, Arak T, Vilo J (2011) g:Profiler—a web server for functional interpretation of gene lists (2011 update). Nucleic Acids Res 39:W307–W315
23. Bauer S, Grossmann S, Vingron M et al (2008) Ontologizer 2.0--a multifunctional tool for GO term enrichment analysis and data exploration. Bioinformatics 24:1650–1651
24. Bauer S, Gagneur J, Robinson PN (2010) GOing Bayesian: model-based gene set analysis of genome-scale data. Nucleic Acids Res 38:3523–3532
25. Kolde R, Vilo J (2015) GOsummaries: an R Package for visual functional annotation of experimental data. F1000Research 4:574
26. Fortino V, Alenius H, Greco D (2015) BACA: bubble chArt to compare annotations. BMC Bioinformatics 16:37
27. Kobayashi SD, Braughton KR, Palazzolo-Ballance AM et al (2010) Rapid neutrophil destruction following phagocytosis of Staphylococcus aureus. J Innate Immun 2:560–575
28. Petryszak R, Burdett T, Fiorelli B et al (2014) Expression Atlas update—a database of gene and transcript expression from microarray- and sequencing-based functional genomics experiments. Nucleic Acids Res 42:D926–D932
29. Weinstein JN, Myers TG, O'Connor PM et al (1997) An information-intensive approach to

the molecular pharmacology of cancer. Science 275:343–349

30. Merico D, Gfeller D, Bader GD (2009) How to visually interpret biological data using networks. Nat Biotechnol 27:921–924
31. Lord PW, Stevens RD, Brass A et al (2003) Investigating semantic similarity measures across the Gene Ontology: the relationship between sequence and annotation. Bioinformatics 19:1275–1283
32. Dimmer EC, Huntley RP, Alam-Faruque Y et al (2012) The UniProt-GO Annotation database in 2011. Nucleic Acids Res 40:D565–D570
33. Schlicker A, Domingues FS, Rahnenführer J et al (2006) A new measure for functional similarity of gene products based on Gene Ontology. BMC Bioinformatics 7:302
34. Mazandu GK, Mulder NJ (2013) Information content-based gene ontology semantic similarity approaches: toward a unified framework theory. BioMed Res Int 2013:292063
35. Wickham H (2009) ggplot2: elegant graphics for data analysis
36. Baehrecke EH, Dang N, Babaria K et al (2004) Visualization and analysis of microarray and gene ontology data with treemaps. BMC Bioinformatics 5:84
37. Smith B, Ashburner M, Rosse C et al (2007) The OBO Foundry: coordinated evolution of ontologies to support biomedical data integration. Nat Biotechnol 25:1251–1255
38. Furnham N (2016) Complementary sources of protein functional information: the far side of GO. In: Dessimoz C, Škunca N (eds) The gene ontology handbook. Methods in molecular biology, vol 1446. Humana Press. Chapter 19

15

Gene-Category Analysis

Sebastian Bauer

Abstract

Gene-category analysis is one important knowledge integration approach in biomedical sciences that combines knowledge bases such as Gene Ontology with lists of genes or their products, which are often the result of *high-throughput* experiments, gained from either wet-lab or synthetic experiments. In this chapter, we will motivate this class of analyses and describe an often used variant that is based on Fisher's exact test. We show that this approach has some problems in the context of Gene Ontology of which users should be aware. We then describe some more recent algorithms that try to address some of the shortcomings of the standard approach.

Key words Enrichment, Overrepresentation, Knowledge integration, Fisher's exact text, Gene propagation problem

1 Introduction

The result of biological *high-throughput* methods is often a list consisting of several hundreds of biological entities, which are in case of gene expression profiling experiments identifiers of genes or their products. As a biological entity may have different context-specific functions, it is difficult for humans to interpret the outcome of an experiment on the basis of such a list. Computational approaches to access the biological knowledge about features of biological entities therefore play an important part in the successful realization of research based on high-throughput experiments. A practical way to address the question of *what is going on?* is to perform a gene-category analysis, i.e., to ask whether these responder genes share some biological features that distinguish them among the set of all genes tested in the experiment.

First of all, gene-category analysis involves a list of gene categories, in which genes with similar features are grouped together. The exact definition of the attribute *similar* depends on the provider of the categories. For instance, if Gene Ontology is the choice, then genes usually are grouped according to the terms, to which they are annotated. Another scheme is the KEGG database [1], in which genes are grouped according to the pathways in which they are involved. The second ingredient is a statistical method for identifying the really interesting categories.

In this chapter, we introduce some commonly used approaches for gene-category analysis. Throughout the remainder of this chapter, we refer to the set of items, which a study could possibly select, as the *population set*. We denote this set by the uppercase letter M while the size of the set, or its *cardinality*, is identified by its lowercase variant m. If, for example, a microarray experiment is conducted, the population set will comprise all genes whose expression can be measured with the microarray chip. The actual outcome of the study is referred to as the *study set*. It is denoted by N and has the cardinality n. In the microarray scenario the study set could consist of all genes that were detected to be differentially expressed.

2 Fisher's Exact Test

One approach for gene-category analysis is to cast the problem as a statistical test. For this purpose, the study set is assumed to be a random sample that is obtained by drawing n items without replacement from the population. The population is dichotomic as the items can be characterized according to whether they are annotated to term t or not. In particular, the set M_t with cardinality m_t constitutes all items that are annotated to t. Denote the random variable that describes the number of items of the study set that are annotated to t in this random sample as X_t. The hypergeometric distribution applies to X_t, and the probably of observing exactly k items annotated to t, i.e., $P(X_t = k)$ is specified by

# of ways of choosing the remaining $n - k$ items that are not annotated to t

# of ways of choosing k items among all items annotated to t

$$X_t \sim h(k|m; mt; n) := P(X_t = k) = \frac{\binom{m_t}{k}\binom{m - m_t}{n - k}}{\binom{m}{n}}.$$

# of ways of choosing n items among m

Furthermore, the set of items that are annotated to t and members of the study set are denoted by N_t with cardinality n_t. The objective is to assess whether the study set is enriched for term t, i.e., whether the observed n^t is higher than one would expect. This forms the alternative hypothesis H_1 of the statistical test. The null hypothesis H_0 in this case is that there is no positive association between the observed occurrence of the items in the study set and the annotations of the items to the term t. Thus, the proportion of items annotated to term t is approximatively identical for the study set and the population set. In order to be able to reject H_0 in support of H_1 we conduct a one-tailed test, in which we ask for the probability of the event that we see n_t or more annotated items given that H_0 is true:

$$p_t^{\text{tft}} = P(X_t \geq n_t \mid H_0) = \sum_{k=n_t}^{\min(m_t, m)} \frac{\binom{m_t}{k}\binom{m-m_t}{n-k}}{\binom{m}{n}}. \quad (1)$$

If the probability obtained by this equation[1] is below a certain significance level α, e.g., $\alpha < 0.05$, we reject H_0 in favor of H_1. In that case, the tested term t is regarded as an interesting term that contributes to the characterization of the study set.

Example 2.1. *Suppose that we are given a population of $m = 18$ genes, of which $m_t = 4$ genes are annotated to a term t. The outcome of an experiment yields a study set of 5 differentially expressed genes. A total of $n_t = 3$ genes from the genes of the study set are annotated to t. Figure 1 illustrates the participating sets and how they are related to one another in that particular situation.*

In order to check whether term t can be used to characterize the experiment, we ask whether term t is overrepresented in the study set. The application of Eq. 1 yields a p-value for t

$$p_t^{\text{tft}} = P(X_t \geq 3 \mid H_0) = \frac{\binom{4}{3}\binom{14}{2}}{\binom{18}{5}} + \frac{\binom{4}{4}\binom{14}{1}}{\binom{18}{5}} = 0.044.$$

Thus, the null hypothesis is rejected and the term is said to be overrepresented among the differentially expressed genes and is thus likely to reflect an association between the term and the experiment.

3 Multiple Testing Problem

In hypothesis-generating studies it is a priori not clear, which terms should be tested. Therefore, the procedure is not only conducted using a single term but also applied to many, often all terms that Gene Ontology provides and to which at least one gene is annotated. The result of the entire analysis is then a list of terms that were found to be significant. This, however, implies that the number of false-positive terms is high.

[1] The superscript *tft* in p_t^{tft} stands for *term-for-term*. It allows to distinguish this p-value with other measures that are described later.

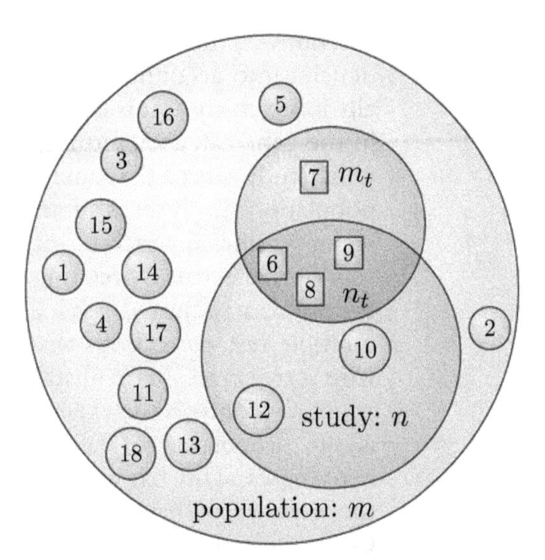

Fig. 1 Sets and their relations in the standard approach. In this example the population consists of $m = 18$ genes and $n = 5$ of them are part of the study set. Exactly $m_t = 4$ genes of the population are annotated to term t. This term has $n_t = 3$ genes in common with the study set. The null hypothesis of the standard approach (*term-for-term*) is that there is no association between the number of genes that are in the study set and the number of genes that are annotated to the term t, i.e., the study set is a random sample of the population set. We therefore would expect that it contains the same proportion of annotated terms as the population set does. The probability under the null hypothesis of the event to see at least *nt* genes can be assessed via Eq. 1.

To see this, suppose that there are T tests to be performed. We assume that the null hypothesis is true for all of those tests. Before its actual determination, any p-value can be considered as a random variable as well, for which $P(p \leq \alpha \mid H_0) \leq \alpha$ holds [2]. This implies that it can be expected that $\alpha \times T$ tests lead to the rejection of a null hypothesis although it is true.

Example 3.1. *If there are 10,000 null hypotheses that are true and all of them are tested, then we expect that we reject the null hypotheses for about 500 tests. Obviously, describing the result of experiment with 500 random terms is not useful.*

Therefore, the result of a term enrichment analysis shall be further subjected to a multiple test correction. The most simple is the Bonferroni correction [3]. Here, each p-value is simply multiplied by the number of tests saturated at a value of 1.0. Bonferroni controls the so-called family-wise error rate, which is the probability of making one or more false discoveries. It is a very conservative approach because it handles all p-values as independent. But as we see later, this is not a typical case of gene-category analysis, so this approach often goes along with a reduced statistical power.

In contrast, the Westfall–Young [4] procedure also takes dependencies into account. This correction, however, is computationally more costly as it is based on resampling schemes. In particular in the gene category setting, this scheme involves randomly sampling study sets of the same size as the original study set from the population. Each set is subjected to the test procedure yielding a set of p-values for each term, also referred to as the *null distribution* of that term. By relating the original p-value to the null distribution, an adjusted p-value is derived. There are other types of multiple test corrections that do not aim to control the family-wise error rate. For instance, the Benjamini–Hochberg [5] approach controls the expected false discovery rate (FDR), which is the proportion of false discoveries among all rejected null hypotheses. This has a positive effect on the statistical power at the expense of having less strict control over false discoveries. Controlling the FDR is considered by the American Physiological Society as "the best practical solution to the problem of multiple comparisons" [6].

Note that less conservative corrections usually yield a higher amount of significant terms, which may be not desirable after all. In the following section, we further explore the structural origin of the correlations of the p-values in the setting of enrichment tests for ontology terms.

4 Gene Propagation

While the application of multiple testing correction aims to reduce the number of false-positives in a rather universal manner, one can also try to tackle the problem at a more basic level. The root of the problem is that if a term shares genes with a second term, and one of the terms is overrepresented, then it is not too surprising that the other term is also detected as overrepresented.

That the gene sharing of terms of an ontology is more a rule than an exception can be deduced from the principles of how ontologies are designed. Within an ontology, terms describe concepts of a domain that can be related to other terms by various types of relationships. The most prominent relationship thereby is the *is a* relationship, which effectively propagates the membership of the subject (source) of the relationship to the object (destination). That means, if a term T_1 is related to a term T_2 by the *is a* relationship, and a gene is annotated to T_1, then it is implicitly annotated also to term T_2 (*see* Chap. 1 [7]). In the context of GO overrepresentation analysis, we refer to this as the *gene propagation problem.*[2]

[2] Note that in addition to this gene sharing that is due to the graph structure of the ontology, also unrelated terms can be annotated to similar sets of genes, for instance, if the same gene plays a role in distinct biological processes.

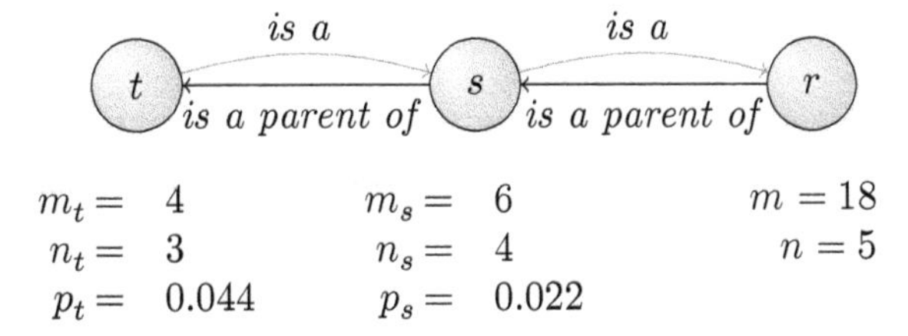

$m_t = 4$ $m_s = 6$ $m = 18$
$n_t = 3$ $n_s = 4$ $n = 5$
$p_t = 0.044$ $p_s = 0.022$

Fig. 2 Extended example with three terms. This depicts the situation of Example 2.1 with two more terms. Term *t is a s* and therefore *s is a parent of t*. Term *r* is the root of the ontology. It is the only parent of *s*. As indicated in the last row, the procedure based on Fisher's exact test determines a *p*-value below 0.05 for both terms. Thus, both terms will be considered as a meaningful summary of the underlying experiment.

Example 4.1 (Continuation of Example 2.1). *There is another term s, which is the only parent of t. For s we know that* $m_s = 6$ *and* $n_s = 4$. *Figure 2 shows this structure graphically. There, it is also indicated that the p-values of terms t and s are 0.044 and 0.022, respectively, which means that both terms are considered as significant for* $\alpha < 0.05$ *if no multiple test correction is performed. Obviously, both terms share the majority of items that are also part of the study set. One can argue that the fact that term t is identified as overrepresented is a consequence of the fact that s is overrepresented.*

A simple synthetic experiment, in which a term will be artificially overrepresented, demonstrates the extent of the problem. Let's select the term *localization* for this purpose. We create a study set that consists of all genes that are annotated to that term with probability 0.8. This corresponds to false-negative rate $\beta = 0.2$. Furthermore, to introduce some background noise, each gene that is not annotated to the term is added to that study set with a false-positive rate of $\alpha = 0.1$. In this example, the procedure yields a set of 1542 genes. For each considered term, this set is subjected to Fisher's exact test resulting in a list of 4549 *p*-values[3]. Finally, the *p*-values are adjusted using the Bonferroni correction.

The analysis correctly identifies the term *localization* as significantly enriched. In addition to that, it identifies 275 other terms as significantly enriched. In particular, 6 of the 6 children, to which at least one gene is annotated, are significant. Among the 681 possible descendants of *localization*, we find 172 significant ones. These figures suggest that descendants come up only because their annotations converge in the term *localization*. Although, in the statistical sense, this is a correct result, it is not desirable to use that huge amount of terms to characterize the study set, especially as it is sufficient to use the term *localization* for this purpose, and what is

[3] This corresponds to the number of terms from the *biological process* subontology that are annotated by at least one gene.

more, the result suggests a specificity that we did not put in there. It makes sense to consider each of the additional 275 significant terms as a false-positive and in the next sections we will briefly describe methods that attempt to reduce that number.

5 Parent–Child Approach

The *parent–child* approach [8] is still based on Fisher's exact test, but the probability of t being overrepresented is conditioned on properties of the parental terms. In the following, let $\mathrm{pa}(t)$ be the set of parents of term t, which are, for instance, those terms, to which t is connected by a *is a* relation. In order to introduce the principal ideas of the *parent–child* approaches, we initially assume that there is only a single parent of t, i.e., $\mathrm{pa}(t) = \{s\}$.

Instead of drawing the items from the population M, items will be drawn just from the set of items that are annotated to the parent of t, which is written as $M_{\mathrm{pa}(t)}$ and whose size is $m_{\mathrm{pa}(t)}$. This consideration yields the following equation:

$$P(X_t = k \mid \mathrm{pa}(t)) = \frac{\binom{m_t}{k}\binom{m_{\mathrm{pa}(t)} - m_t}{n_{\mathrm{pa}(t)} - k}}{\binom{m_{\mathrm{pa}(t)}}{n_{\mathrm{pa}(t)}}}. \quad (2)$$

The right part of Fig. 3 shows the setting of the *parent–child* approaches. Effectively, in the *parent–child* approaches, we change the population that underlies Fisher's exact test to the items annotated to the parents. Obviously, this also alters the involved sets for the study set. As previously, we ask for the probability of seeing the observed number of items or a more extreme event:

$$p_t^{\mathrm{pc}} = P(X_t \geq n_t \mid H_0) = \sum_{k=n_t}^{\min(m_t, m_{\mathrm{pa}(t)})} \frac{\binom{m_t}{k}\binom{m_{\mathrm{pa}(t)} - m_t}{n_{\mathrm{pa}(t)} - k}}{\binom{m_{\mathrm{pa}(t)}}{n_{\mathrm{pa}(t)}}}. \quad (3)$$

Example 5.1 (Continuation of Example 4.1). *As shown in Fig. 2, the parent of term s is the root r of the ontology, which is always annotated to all genes of the population. Therefore, the p-value for s is the same for previous approach and for* parent–child *approach, i.e.,* $p_s^{\mathrm{pc}} = p_s = 0.22$ *However, for term t, Eq. 3 yields*

$$p_t^{\mathrm{pc}} = P(X_t \geq n_t \mid H_0) = \frac{\binom{4}{3}\binom{2}{1}}{\binom{6}{4}} + \frac{\binom{4}{4}\binom{2}{0}}{\binom{6}{4}} = 0.6.$$

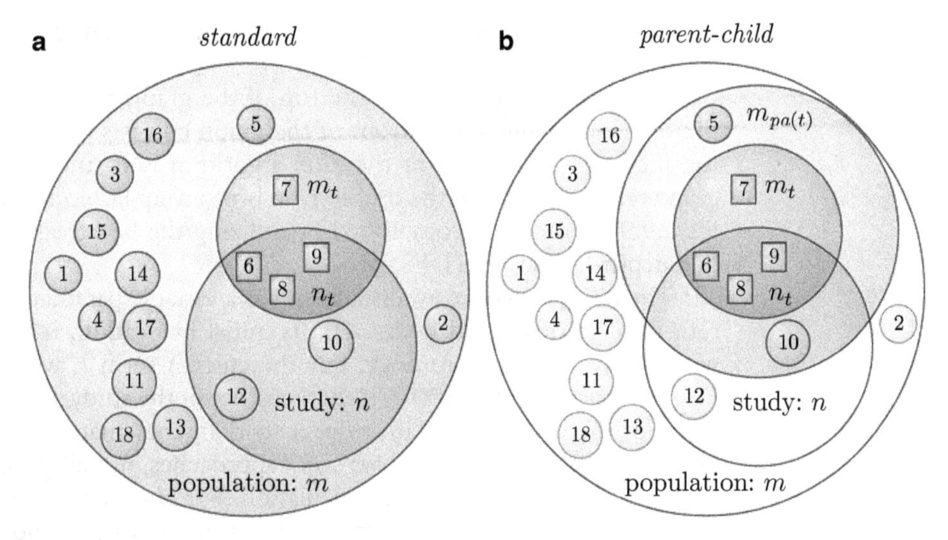

Fig. 3 Sets and their relations in the *parent–child* approaches. Part (**a**) depicts the model of the *term-for-term* approach as it was shown in Fig. 1. This is contrasted in part (**b**) with the model of the *parent–child* approaches. In this approach, we shift the focus to a smaller set of genes, for instance to the genes that are annotated to at least one of the parents of term t. In this particular situation it is the set whose size is $m_{\text{pa}(t)} = 6$ with $\text{pa}(t) = \{s\}$ following Example 4.1. Genes that are not part of this set do not contribute to the calculation. This has an effect on the involved proportions, and thus on the outcome of the test. Effectively, for each term, we alter the population of the association test. Eq. 2 quantifies the probability.

Thus, the null hypothesis for term t is not rejected, which is in contrast to the result of the previous approach. Given the initial observations that the study set is already skewed to the parent s of t makes the enrichment of term t less surprising, which the parent–child *approaches reflect by returning a higher p-value.*

If term t has more than one parent term, then it is not immediately apparent how to calculate $m_{\text{pa}}(t)$ and the observation $n_{\text{pa}}(t)$ in Eqs. 2 and 3. In Grossmann et al. [8] we examined two variants in detail, the union and the intersection of genes that are annotated to each of the parents.

6 Topology-Based Algorithms

Alexa et al. devised another method to address the *gene propagation problem*. The authors propose calculating a score for the term that depends on the relevance of the children of the term [9]. They argue that capturing the meaning in that way is biologically more interesting as the definitions of children are more specific. Following this argumentation, the authors formulated two concrete algorithms that try to provide a more suitable, i.e., less correlated, distribution of terms that get flagged as important. While the first approach which they called the *elim*-algorithm strictly favors significance of the most specific levels of the GO graph, their second algorithm called *weight* relaxes this restriction such that terms that are most significant are favored.

As before, we understand the top of the graph as the root of the ontology, while the bottom of the graph consists of the most specific terms. The idea of the *elim* algorithm is to traverse the graph representation of the ontology in bottom-up fashion, which, for instance, can be accomplished by utilizing the backtrack phase of a depth-first search (DFS) [10].

The *elim* procedure awaits a term t as a variable parameter and returns a set of flagged genes. On its initial invocation, it begins with the root of the ontology. For the current term t, we apply Fisher's exact test in order to relate the genes of the study set to the genes of the population with respect to the genes that are annotated to term t. As in the *parent–child* approaches, not all genes of the study set contribute to the calculation. For *elim*, a set of previously determined genes is subtracted from the set of the study set before the calculation for p_t is carried out. This set is constructed by recursively applying the *elim* procedure for all children of t and taking the union of the result. If p_t is significant, we add all genes of t to the set of flagged genes. Finally, we return the set of flagged genes to the caller. Note that when the DFS reaches a leaf node of the ontology, Fisher's exact test is performed exactly as in the standard approach.

Obviously, the complexity of the algorithm is the same as the complexity of a depth-search algorithm if we assume that the number of genes that are annotated to a term is constant. Note in the original publication of the *elim*, the algorithm was based on an iteration over the levels of the GO DAG, which partitions the nodes according to their longest distance to the root. The algorithm as outlined here yields an equivalent result without the need to explicitly keep track of the DAG levels.

Example 6.1 (Continuation of Example 5.1). *The p-value of term t matches the p-value of term t of the standard approach, i.e.,* $p_t^{\text{elim}} = p_t^{\text{tft}} = 0.044$. *As this is a significant result, at least, if correction for multiple testing is omitted, all four genes that are annotated to t are removed in the consideration of upper terms, i.e., we assume that those four genes are not annotated to them. This leaves two genes for the computation of term s, of which only one is member of the study set (Fig. 3b). With* $m_s = 2$, $n_s = 1$, *and the rest as before, Eq. 1 yields*

$$p_s^{\text{elim}} = P(X_s \geq 1 \mid H_0) = \frac{\binom{2}{1}\binom{16}{4}}{\binom{18}{5}} + \frac{\binom{2}{2}\binom{16}{3}}{\binom{18}{5}} = 0.49.$$

Hence, the elim *method doesn't report term s as important.*

An equivalent characterization of the *elim* method is the following: If a term t is identified as significant, all genes that are annotated to t are no longer considered in the computation of the relevance of the ancestors of t. As it was discussed in Example 2.1 at page 2.1 and as can also be seen in Fig. 2, the *term-for-term* approach assigns term s a lower p-value than it does for term t. One may conclude that it is more appropriate to take term s than to take term t in order to provide a compact description of the study set. However, in Example 13.6.1 we saw that the application of the *elim* method results in usage of term t to describe the outcome, which is contrary to that conclusion.

This concern is addressed by *weight* method. It compares significance scores of a family terms (a parent and its child) to identify the locally most significant terms and down-weight genes in less significant neighbors. This effectively decorrelates the p-values of the related terms such that their differences are enforced while the existence of the most significant terms is still maintained.

7 Model-Based Approaches

The previously described procedures that address gene propagation problem have in common that they successively test overrepresentation for each of the terms. They all use some form of the Fisher's exact test. In contrast to this, model-based gene set analysis (MGSA) models the gene response in a genome-wide experiment as the result of an activation of a number of terms [11].[4]

The approach is based on a model that can nicely be expressed using a Bayesian network with three layers of Boolean random variables. The *term layer* consists of m Boolean nodes corresponding to m terms of the ontology. A term can be *active* or *inactive*. A parameter p, usually much less than 0.5, represents the prior probability of a term being active. The hidden layer contains n Boolean nodes representing the n hidden state of the genes. The hidden state of a gene is a consequence of the states the terms to which the gene is annotated: The gene is *on* if and only if at least one term to which the gene is annotated is active, otherwise it is *off*. The third layer, the *observed layer*, contains Boolean nodes reflecting the experimentally observed state of all genes. For instance, in the setting of a microarray experiment, the *on* state would correspond to differential expression, and the *off* state would correspond to a lack of differential expression of a gene. The observed gene state depends on the corresponding hidden gene state in a one-to-one fashion with a false-positive (α) and false-negative rates (β) that is identical and independent for all genes. A simple instance of the model is depicted in Fig. 4.

[4] We use the word *term* here because we primarily work with GO, but the method can be applied to any other structured or unstructured vocabulary.

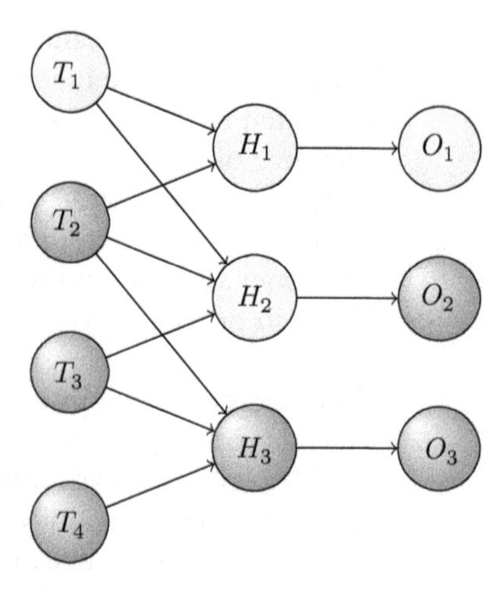

Fig. 4 The graphical representation of an MGSA network. An example structure for four terms and three genes with a possible realizations is displayed. Terms (T_i) that constitute the first layer can be either *active* (*light*) or *inactive* (*dark*). Terms that are *active* enable the hidden state (H_j) of all genes annotated to them, the other genes remaining *off*. The observed states (O_j) of the genes are noisy observations of their true hidden state. In this example, the observed states for gene 1 and 3 match the hidden state while for some unknown reasons the measurement of gene 2 doesn't correspond to the hidden state. It's a false-negative.

The model describes how the activity of terms leads to the observed stats of genes. This, however, is not the direction we are interested in. We are interested in the set of terms that explain the experimentally obtained data best, and the mathematical tool that can be applied to and such sets is probabilistic inference. The optimization problem that finds the term state configuration that explains the observed gene pattern best is NP-hard [12]. However, it is easily possible to find nearby solutions by sampling from the state space. This procedure additionally allows to determine the so-called marginal probability for each term, which is a measure how good the particular term will explain the observed genes with respect to all the other terms. The value ranges between 0 and 1 with 0 being the lowest possible support and 1 being the best possible support for a term. As all terms compete with one another, the inference takes dependencies both due to gene propagation and due to similarity of annotations into account. For example, if two unrelated terms are annotated to the same set of genes that matches the observation, the marginal probability for both terms will be 0.5. Consequently, it is advisable to run MGSA for each of the subontologies separately as they are designed to express orthogonal features.

8 Gene Set Enrichment Analysis

In addition to approaches that take a fixed subset of the population as input, procedures that take the measurements of the genes into account are also widely in use. This is attractive as it frees the investigator from the need to define a sometimes arbitrary cutoff that is used to construct the study set.

A first version of the so-called Gene Set Enrichment Analysis (GSEA) that received much attention of the scientific community was published by Mootha et al. [13]. In this approach, genes are ranked according to an interesting feature (e.g., the difference of the mean of their expression values for two experimental conditions). The null hypothesis is that the genes of the interesting set (e.g., genes annotated to a term) have no association with that list, in which case they would be randomly ordered. The alternative hypothesis is that the genes of the interesting set have an association. For instance, if the genes of the set are grouped together on the top of the list, we would tend to believe that there is such an association.

To capture the association via statistical means, the authors proposed a normalized Kolmogorov–Smirnov (KS) test statistic. Let $r_i \in M$ be the gene of the population M that has rank i in the gene list that is sorted according to the interesting gene feature. Using the previously established notation, i.e., that m is the total number of genes and N_t is the set of cardinality n_t that contains only genes that are annotated to t, the score is defined as:

$$ES(N_t) = \max_{i\in\{1,\ldots,m\}} \sum_{j=1}^{i} X_j \text{ with } X_j = \begin{cases} -\sqrt{\dfrac{n_t}{m-n_t}}, & \text{if } r_i \notin N_t \\ \sqrt{\dfrac{m-n_t}{n_t}}, & \text{otherwise} \end{cases}$$

Thus, the score is the maximum of a running sum that is increased if the gene is annotated to t and decreased if the gene is not annotated to t. In order to check if the obtained score is significant, the calculation is repeated for k randomly chosen sets $N_t^1, \ldots, N_{tk}$, which all are subsets of M with size n_t. The p-value for a term t is calculated as

$$p_t = \frac{\left|\{i \mid ES(N_t^i) \geq ES(N_t)\}\right|}{k}.$$

The GSEA method went a slight revision Subramanian et al. [14], where ad-hoc modifications are implemented that are supposed to countervail the well-known lack of sensitivity of the KS test [15, 16].

9 Software

Gene-category analysis is a very prominent use case of Gene Ontology. It shouldn't come as a surprise that users can choose among a variety of software implementations that will perform this sort of analysis. For instance, current version of the web site of Gene Ontology Consortium (geneontology.org) provides access to the method of the basic Fisher's exact test directly on the front page. There are also graphical tools that integrate into existing frameworks such as *BiNGO* [17], standalone graphical clients such as *Ontologizer*[5] [18] or packages for Bioconductor such as *topGo* [19], *mgsa* [20], or *gCMAP* [21], just to name a few of them.

10 Exercises

1. Repeat the random experiment outlined in the text that was used to show the influence of the gene propagation. When doing this in R/Bioconductor, it is advisable to use the *GO.db* and *org.Sc.sgd.db* packages that provide the structure and the annotations. The calculation involving the hypergeometric distribution can be expressed directly in R using dhyper and phyper. Now repeat this experiment with other approaches based on study sets that were outlined in this chapter and compare the results. For the topology-based algorithms the *topGo* package can be used and for the model-based approach the *mgsa* package is well suited.
2. Apply the approach now to an arbitrary example or on real world data. Compare the results.

[5] http://ontologizer.de

References

1. Kanehisa M, Goto S (2000) KEGG: Kyoto encyclopedia of genes and genomes. Nucleic Acids Res 28(1):27–30
2. Ewens WJ, Grant GR (2005) Statistical methods in bioinformatics: an introduction, 2nd edn. Springer, Berlin. ISBN 978-0387400822
3. Abdi H (2007) Bonferroni and Sidak corrections for multiple comparisons. Sage, Thousand Oaks, CA
4. Westfall PH, Young SS (1993) Resampling-based multiple testing: examples and methods for P-value adjustment. Wiley, London. ISBN 978-0471557616
5. Benjamini Y, Hochberg Y (1995) Controlling the false discovery rate: a practical and powerful approach to multiple testing. J R Stat Soc Ser B 57:289–300
6. Curran-Everett D, Benos DJ (2004) Guidelines for reporting statistics in journals published by the American Physiological Society. Adv Physiol Educ 28:85–87
7. Hastings J (2016) Primer on ontologies. In: Dessimoz C, Škunca N (eds) The gene ontology handbook. Methods in molecular biology, vol 1446. Humana Press. Chapter 1
8. Grossmann S, Bauer S, Robinson PN, Vingron M (2007) Improved detection of overrepresentation of Gene-Ontology annotations with parent child analysis. Bioinformatics 23:3024–3031
9. Alexa A, Rahnenführer J, Lengauer T (2006) Improved scoring of functional groups from gene expression data by decorrelating GO graph structure. Bioinformatics 22(13):1600–1607. doi:10.1093/bioinformatics/btl140
10. Cormen TH, Leiserson CE, Rivest RL, Stein C (2001) Introduction to algorithms, 2nd edn. MIT Press, Cambridge, MA. ISBN 978-0262531962
11. Bauer S, Gagneur J, Robinson PN (2010) GOing Bayesian: model-based gene set analysis of genome-scale data. Nucleic Acids Res 38(11):3523–3532
12. Bauer S (2012) Algorithms for knowledge integration in biomedical sciences. PhD thesis
13. Mootha VK, Lindgren CM, Eriksson K-F, Subramanian A, Sihag S, Lehar J, Puigserver P, Carlsson E, Ridderstråle M, Laurila E, Houstis N, Daly MJ, Patterson N, Mesirov JP, Golub TR, Tamayo P, Spiegelman B, Lander ES, Hirschhorn JN, Altshuler D, Groop LC (2003) PGC-1α-responsive genes involved in oxidative phosphorylation are coordinately downregulated in human diabetes. Nat Genet 34(3):267–273. doi:10.1038/ng1180
14. Subramanian A, Tamayo P, Mootha VK, Mukherjee S, Ebert BL, Gillette MA, Paulovich A, Pomeroy SL, Golub TR, Lander ES, Mesirov JP (2005) Gene set enrichment analysis: a knowledge-based approach for interpreting genome-wide expression profiles. Proc Natl Acad Sci USA 102(43):15545–15550. doi:10.1073/pnas.0506580102
15. Mason DM, Schuenemeyer JH (1983) A modified Kolmogorov-Smirnov test sensitive to tail alternatives. Ann Stat 11(3):933–946
16. Irizarry RA, Wang C, Zhou Y, Speed TP (2009) Gene set enrichment analysis made simple. Stat Methods Med Res 18(6):565–575. ISSN 1477-0334
17. Maere S, Heymans K, Kuiper M (2005) Bingo: a cytoscape plugin to assess overrepresentation of gene ontology categories in biological networks. Bioinformatics 21:3448–3449
18. Bauer S, Grossmann S, Vingron M, Robinson PN (2008) Ontologizer 2.0–a multifunctional tool for go term enrichment analysis and data exploration. Bioinformatics 24(14):1650–1651. doi:10.1093/bioinformatics/btn250
19. Alexa A, Rahnenführer J (2010) topGO: enrichment analysis for Gene Ontology. R package version 2.22.0
20. Bauer S, Robinson NP, Gagneur J (2011) Model-based Gene Set Analysis for Bioconductor. Bioinformatics 27
21. Sandmann T, Kummerfeld SK, Gentleman R, Bourgon R (2014) gcmap: user-friendly connectivity mapping with r. Bioinformatics 30(1):127–128

16

Get GO! Retrieving GO Data using AmiGO, QuickGO, API, Files and Tools

Monica Munoz-Torres and Seth Carbon

Abstract

The Gene Ontology Consortium (GOC) produces a wealth of resources widely used throughout the scientific community. In this chapter, we discuss the different ways in which researchers can access the resources of the GOC. We here share details about the mechanics of obtaining GO annotations, both by manually browsing, querying, and downloading data from the GO website, as well as computationally accessing the resources from the command line, including the ability to restrict the data being retrieved to subsets with only certain attributes.

Key words Gene ontology, Ontology, Annotation resources, Annotation, Genomics, Transcriptomics, Bioinformatics, Biocuration, Curation, Access, AmiGO, QuickGO

1 Introduction

The efforts of the Gene Ontology Consortium (GOC) are focused on three major subjects: (1) the development and maintenance of the ontologies; (2) the annotation of gene products, which includes making associations between the ontologies and the genes and gene products in all collaborating databases; and (3) the development of tools that facilitate the creation, maintenance, and use of the ontologies. This chapter is focused on the mechanics of obtaining GO annotations, both directly and computationally, including the ability to restrict the data being retrieved to subsets with only certain attributes.

GO data is the culmination of various forms of curation, made accessible through a variety of interfaces and downloadable in different forms, depending on your intended use. Because the data and software landscape are constantly changing, it is hard to cover with any permanence the best way to access the data; this inherent limitation should be kept in mind as we navigate through this section. This chapter is intended as an overview of the different ways users can access GO data (via web portals, downloadable files, and API) a quick description of basic software used by GO, and as a reference for where to find more detailed and up-to-date information about these subjects.

2 Web Interfaces to Access the GO

This section covers the online interfaces for accessing and interacting with the data using standard web browsers. Most consumers of the GO can make use of data browsers such as AmiGO, QuickGO, and data browsers embedded within more specific databases.

2.1 AmiGO

AmiGO ([1] http://amigo.geneontology.org; Fig. 1a) is the official web-based open-source tool for querying, browsing, and visualizing the Gene Ontology and annotations collected from the MODs (model organism databases), UniProtKB, and other sources (complete list of member institutions currently contributing to the GOC at http://geneontology.org/page/go-consortium-contributors-list). Notable features include: basic searching, browsing, the ability to download custom data sets, and a common question "wizard" interface. Recent changes have brought improvements both in speed and the variety of search modes, as well as the availability of additional data types, such as the display of annotation extensions (*see* Chap. 17 [2]) and display of protein forms (splice variants and proteins with post translational modifications). More details about the latest improvements on the AmiGO browser can also be found at GOC—Munoz-Torres (CA), 2015 [3].

2.2 QuickGO

The Gene Ontology Annotation (GOA) project at the European Molecular Biology Laboratory's European Bioinformatics Institute (EMBL-EBI) also makes available the QuickGO browser ([4]; http://www.ebi.ac.uk/QuickGO; Fig. 1b), a web-based tool that allows easy browsing of the Gene Ontology (GO) and all associated electronic and manual GO annotations provided by the GO Consortium annotation groups. Included in its many features are extensive search and filter capabilities for GO annotations, a powerful integrated subset/slim interface, as well as an integrated historical view of the terms. For data consumption, QuickGO provides broad-ranging web services and cart functionality (a way of persisting abstract elements, like term IDs, between parts of the QuickGO web application).

AmiGO and QuickGO make use of the same GO data sets, with somewhat different implementations according to the requirements of funding sources and respective users. AmiGO, in its entirety, is a product of the GO Consortium and is the official channel for dissemination of the GO data sets, adhering to funding recommendations from NHGRI-NIH. QuickGO is produced, managed, and funded by EMBL-EBI; the members of QuickGO's managing team are also members of the GOC.

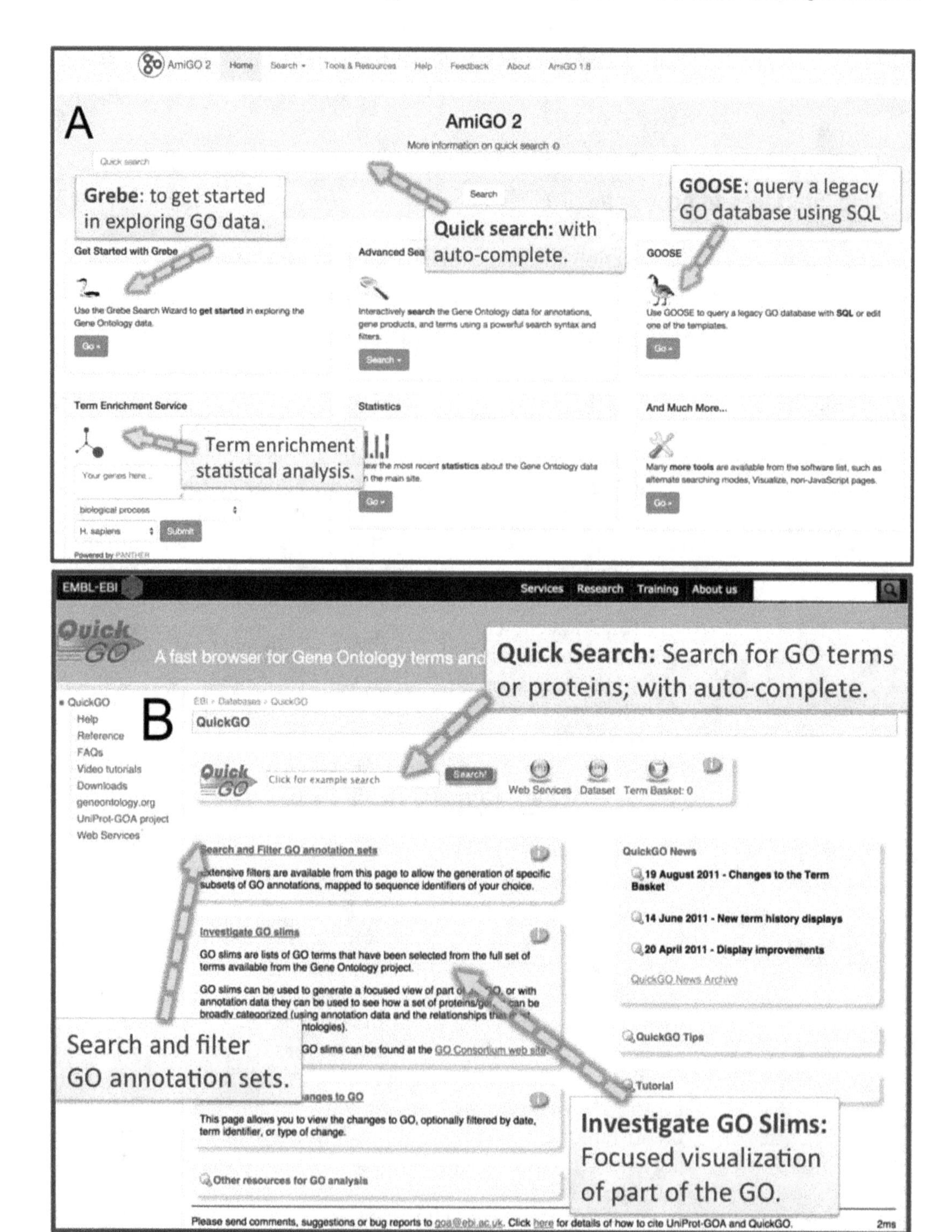

Fig. 1 Landing pages for the AmiGO (**a**) and QuickGO (**b**) browsers. A few features are highlighted for each browser

2.3 Other Browsers

The ontology component of the GO is also searchable and browsable from various third party generic ontology browsers such as OntoBee (http://ontobee.org), the EMBL-EBI Ontology Lookup Service (OLS) (http://www.cbi.ac.uk/ontology-lookup), OLSVis (http://ols.wordvis.com), and BioPortal (http://bioportal.bioontology.org). Each of these systems has their own particular strengths—for example OntoBee is aimed at the semantic web community, and provides the ontology as part of a linked data platform [5], whereas OLSVis is geared towards visualization. However, none of these browsers currently provide access to the annotations.

2.4 Term Enrichment Tool

One of the main uses of the GO is to perform enrichment analysis on gene sets. For example, given a set of genes that are up regulated under certain conditions, an enrichment analysis will find which GO terms are overrepresented (or underrepresented) using the available annotations for that gene set. The GO website offers a service that directly connects users with the enrichment analysis tool from the PANTHER Classification System [6]. The PANTHER database is up-to-date with GO annotations, and their enrichment tool is driven by GO data. Further details about this enrichment tool, as well as a list of supported gene IDs, are available from the PANTHER website at http://www.pantherdb.org/ and at http://www.pantherdb.org/tips/tips_batchIdSearch_supportedId.jsp. More information on enrichment analysis using the GO is available Chap. 13 [7] on "*Gene-Category Analysis.*"

2.5 A Simple Example of Data Exploration Using AmiGO (See Fig. 2)

To give a concrete example of the type of easy GO data exploration that can be accomplished using a web interface, we here provide an example where a user on the AmiGO annotation search interface (http://amigo.geneontology.org/amigo/search/annotation) is trying to find associations between genes/gene products and epithelial processes, while searching only data outside those available for human, and which have experimental evidence.

The user could:

- Type "*epithel*" into the text filter box (*Free-text filtering*) to the left of the results area.
- Open the "*Taxon*" facet and select the [–] next to "*Homo sapiens.*"
- Open the "*Evidence type*" facet and select the [+] next to "*experimental evidence.*"

The remaining results would fit the initial search criteria. However, suppose that the user wants to further refine their search to strictly look at all GO annotations that are directly or indirectly annotated to the GO term "*epithelial cell differentiation*" (GO:0009913). Following the steps above, they could:

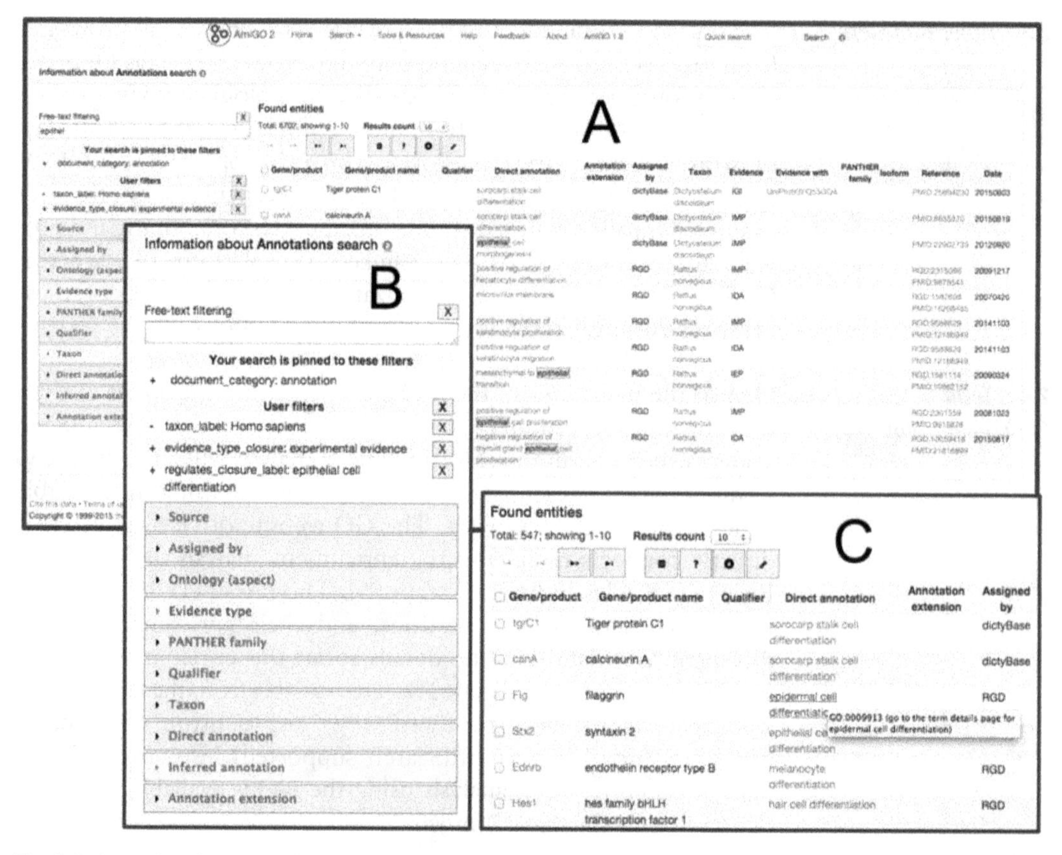

Fig. 2 Data exploration using the AmiGO annotation search interface. All results from this example are listed in *panel* (**a**). (**b**) Shows a detail about the filters applied throughout the search, listed under "**User filters**." An example of the details that appear for each gene or gene product is visible in (**c**): note that the information about the GO term ID for "*epithelial cell differentiation*" (GO:0009913) appears when users hover over the "Direct annotation" details

- Open the "*Inferred annotation*" facet and select the [+] next to "*epithelial cell differentiation*," then
- Remove the text filter by clicking the [x] next to the text entry.

This would leave the user with all GO annotations directly or indirectly annotated with "*epithelial cell differentiation*" (GO:0009913), that are not from human data, and have some kind of experimental evidence associated with them.

3 GO Files: Description and Availability

GO data files contain the current and long-term output of ontology and annotation efforts that are used for exchanging data across various systems. There are several use cases where it may be easier to mine the data directly from the files using a variety of tools. The most commonly used raw data files can be broken down into two categories: ontology and association files.

3.1 Ontology

In the context of GO, ontologies are graph structures comprised of classes for molecular functions, the biological processes they contribute to, the cellular locations where they occur, and the relationships connecting them all, in a species-independent manner [3]. Each term in the GO has defined relationships to one or more other terms in the same domain, and sometimes to other domains. Additional information about ontologies in general is also available from Chap. 1 [8].

GO ontology data are available from the GO website at http://geneontology.org/page/download-ontology. There are three different editions of the GO, in increasing order of complexity: go-basic, go, and go-plus.

go-basic: This basic edition of the GO is filtered such that annotations can be propagated up the graph. The relations included are *is_a*, *part_of*, *regulates*, *negatively_regulates*, and *positively_regulates*. It is important to note that this version excludes relationships that cross the three main GO hierarchies. Many legacy tools that use the GO make these assumptions about the GO, so we make this version available in order to support these tools. This version of the GO ontology is available in OBO format only.

go: This core edition of the GO includes additional relationship types, including some that span the three GO hierarchies, such as *has_part* and *occurs_in*, connecting the otherwise disjoint hierarchies found in **go-basic**. This version of the GO ontology is available in two formats, OBO and OWL-RDF/XML.

go-plus: This is the most expressive edition of the GO; it includes more relationships than **go** and connections to external ontologies, including the Chemical Entities of Biological Interest ontology (ChEBI; [9]), the Uberon anatomy (or stage) ontology [10], and the Plant Ontology for plant structure/stage (PO; [11]). It also includes import modules that are minimal subsets of those ontologies. This allows for cross-ontology queries, such as "*find all genes that perform functions related to the brain*" (e.g., in AmiGO: http://amigo.geneontology.org/amigo/term/UBERON:0000955#display-associations-tab). **go-plus** [12] also includes rules encoding biological constraints, such as the spatial exclusivity between a nucleus and a cytosol. These constraints are used for validation of the ontology and annotations [13]. This version of the GO ontology is available in OWL-RDF/XML.

When working with the ontologies, the official language of the Gene Ontology is the Web Ontology Language, or **OWL**, which is a standard defined by the World Wide Web Consortium (W3C). The GO has approximately 41,000 terms covering over 4 million genes in almost 470,000 species [3]. Its organization goes beyond a simple terminology structured as a directed acyclic graph (DAG), as it consists of over 41,000 classes, but it also includes an import chain that brings in an additional 10,000 classes from additional ontologies ([10] and see "*go-plus*" above). In order to best represent the

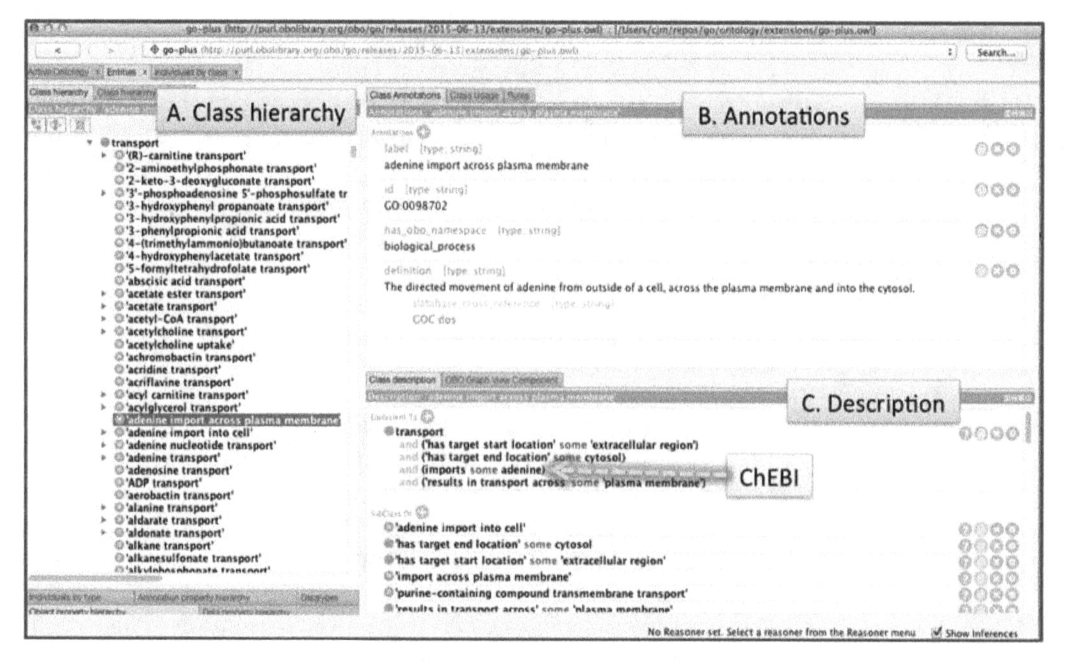

Fig. 3 Visualizing a GO term using Protégé. Protégé displays the details of the term "*adenine import across plasma membrane*" (GO:0098702). The underlying structure of the term is written in the OWL language, which adds flexibility to the expression of associations between genes and gene products and the terms in the ontology, compared to the possibilities offered in OBO. For example, in this term, inter-ontology logical definitions (OWL axioms) coming from the ChEBI ontology [9] are visible; this is not possible to see when visualizing the ontology using OBO

complexity of these classes, along with the approximately 27 million associations that connect them to each molecular entity (genes or gene products), members of the GOC software development team worked on building an axiomatic structure for GO. That is, they assigned logical definitions (known as OWL axioms or self-evidently true statements) to all the classes; the Gene Ontology has been effectively axiomatized, that is, reduced to this system of axioms in OWL, and is highly dependent on the OWL tool stack [10]. Examples of OWL stanzas for terms that are defined by a logical definition in the Gene Ontology are available from GOC—Munoz-Torres (CA), 2015 [3].

A number of tools, frameworks, and software libraries support OWL, including the ontology editor Protégé (http://protege.stanford.edu/; Fig. 3), the Java OWL API, and the OWLTools framework produced by the GO (https://github.com/owlcollab/owltools). Figure 3 shows a GO term visualized using Protégé; its underlying structure is the OWL language. We also make the ontology editions available in OBO Format, which is a simpler format used in many bioinformatics applications (note that "*go-plus*" is not available in OBO format). The two formats can be interconverted using the *Robot* tool produced by the GO Consortium, which can be found at https://github.com/ontodev/robot/.

The GO project is constantly evolving, and it welcomes feedback from all users (*see* below in Subheading 5.3). Research groups may contribute to the GO by either providing suggestions for updating the ontology (e.g., requests for new ontology terms) or by providing annotations. Requests for new synonyms or clarification of textual definitions are also welcomed.

Annotators and other data creators can search whether a term currently exists using the AmiGO browser at http://amigo.geneontology.org/, or may request new ones using either the GO issue tracker on GitHub or TermGenie. TermGenie ([14]; http://termgenie.org) is a web-based tool for requesting new Gene Ontology classes. It also allows for an ontology developer to review all generated terms before they are committed to the ontology. The system makes extensive use of OWL axioms, but can be easily used without understanding these axioms. Users not yet familiar with TermGenie, or whom do not yet have permission to use directly, may submit ontology updates and requests using the GO curator request tracker on GitHub (https://github.com/geneontology/go-ontology/issues), which allows free-text form submissions. For more information on how to best contribute to the GO, please *see* Chap. 7 [15].

3.2 Ontology Subsets

Gene Ontology subsets (also sometimes known as "*slims*") are cut-down versions of the ontologies, containing a reduced number of terms (e.g., species-specific subsets or more generic subsets with "useful" terms in various categories). They give a broad overview of the ontology content without the detail of the specific fine-grained terms. Subsets are particularly useful for giving a summary of the results of GO annotation of a genome, microarray, or cDNA collection when broad classification of gene product function is required. Further information, including Java-based tools and data downloads, is available from the GO website (http://geneontology.org/page/go-slim-and-subset-guide).

3.3 Association Files

The annotation process captures the activities and localization of a gene product using GO terms, providing a reference, and indicating the kind of available evidence in support of the assignment of each term using evidence codes. Currently, the main format for annotation information in the GO is the Gene Association File (GAF, http://geneontology.org/page/go-annotation-file-formats). This is the standardized file format that members of the Consortium use for submitting data. The annotation data is stored in tab-delimited plain text files, where each line in the file represents a single association between a gene product and a GO term, with an evidence code, the reference to support the link between them, and other information. The GAF file format has several different "flavors," with 2.1 being the most current version. Additional details about GAF files is found in Chap. 3 [16].

Recently, the GPAD/GPI files were developed, which are essentially a normalized version of GAF information. These formats are expected to have more prominence in the future, and further details about them can be found on the GO website (http://geneontology.org/page/go-annotation-file-formats).

Because they are tab-delimited text files, both the GAF and GPAD/GPI file formats are very amenable to mining with command line tools. As well, OWLTools can also be used to access this annotation information with operations such as: connecting the annotations to ontology information for exploration and reasoning, OWL translation, validation, taxon checks, and link prediction. More advanced details on this topic are further explained on the OWLTools project wiki (https://github.com/owlcollab/owltools/wiki).

Details on how to make and evaluate GO annotations are discussed in Chap. 4 [17] on "*Best Practices in Manual Annotation with the Gene Ontology*," and in Chap. 8 [18] on "*Evaluating Computational Gene Ontology Annotations.*" Information is also available in the GO Annotation Guide (http://geneontology.org/page/go-annotation-policies); more information on the meaning and use of the evidence codes in support of each annotation can be found on the GO Evidence Codes documentation (http://geneontology.org/page/guide-go-evidence-codes). The GOC is currently transitioning from using evidence codes into implementing the Evidence Ontology (ECO) to describe the evidence in support of each association between a gene product and a GO term. A detailed description of the Evidence Ontology and its use cases is included in Chap. 18 [19] on "*The Evidence and Conclusion Ontology: Supporting Conclusions & Assertions with Evidence.*"

4 Making Your Own Tools

In addition to using off-the-shelf tools provided by the GOC or other users, we also provide libraries and APIs to enable end-users to easily create their own tools for working with and analyzing GO data.

Within the Java/JVM ecosystem, the OWLTools (https://github.com/owlcollab/owltools), and OWL API (https://github.com/owlcs/owlapi) libraries are the primary tools to work with the data. Since OWL is the internal representation format used by the GOC, standard OWL reasoners and tools are all usable with the data. For slightly less general access to the data, the OWLTools(-Core) wrapper library adds numerous helper methods to access OBO-specific fields (i.e., *synonyms*, *alt_ids*), walk graphs, create closures, and other common operations.

On the JavaScript side (both client and server), AmiGO development has produced JavaScript APIs (http://wiki.geneontology.org/index.php/AmiGO_2_Manual:_JavaScript) and widgets (http://

wiki.geneontology.org/index.php/AmiGO_2_Manual:_Widgets) for better access and integration with other tools. Users interested in using the JavaScript API or widgets from AmiGO in their own site should become familiar with the manager and response interfaces, which are the core of the JavaScript interface. An introductory overview of the JavaScript API and widgets, as well as details on implementation engines, the response class, and the configuration class can be also be found on the JavaScript section of the AmiGO Manual, listed above.

As well, AmiGO provides methods for producing incoming searches to allow external sites to link to relevant information. Documentation about these methods can be found at http://wiki.geneontology.org/index.php/AmiGO_2_Manual:_Linking.

5 Additional Information

5.1 Mappings

The GO project provides mappings between GO terms and other key related systems (built for other purposes), such as Enzyme Commission numbers or Kyoto Encyclopedia of Genes and Genomes (KEGG). However, one should be aware that these mappings are neither complete nor exact and should be used with caution. A complete listing of mappings available for the resources of the GOC can be found at http://geneontology.org/page/download-mappings. Additional information about alternative and complementary resources to the GO is available on Chap. 19 [20].

5.2 Legacy Interface for GO

Currently, the AmiGO and QuickGO interfaces have moved away from SQL database derivatives of the data sets. However, to support legacy applications and queries, the GO data is regularly converted into an SQL database (MySQL). These builds can be downloaded and installed on a local machine, or queried remotely using the GO Online SQL/Solr environment (GOOSE; http://amigo.geneontology.org/goose). More information about SQL access, including various downloads and schema information, can be found in the legacy SQL section of the GO website (http://geneontology.org/page/lead-database-guide).

5.3 Help/Troubleshooting Software and Data

In additional to other functions, the GO Helpdesk addresses user queries about the Gene Ontology and related resources. The GO Helpdesk will direct any questions or concerns with GO data, software, or analysis to the appropriate people within the consortium. You can directly contact the GO Helpdesk using the site form (http://geneontology.org/form/contact-go), which will automatically enter your query into an internal tracker to ensure responsiveness.

References

1. Carbon S, Ireland A, Mungall CJ, Shu S, Marshall B, Lewis S, AmiGO Hub, Web Presence Working Group (2009) AmiGO: online access to ontology and annotation data. Bioinformatics 25(2):288–289
2. Huntley RP, Lovering RC (2016) Annotation extensions. In: Dessimoz C, Škunca N (eds) The gene ontology handbook. Methods in molecular biology, vol 1446. Humana Press. Chapter 17
3. Gene Ontology Consortium, Munoz-Torres MC (Corresponding Author) (2015) Gene Ontology Consortium: going forward. Nucleic Acids Res 43(Database issue):D1049–D1056. doi:10.1093/nar/gku1179
4. Binns D, Dimmer E, Huntley R, Barrell D, O'Donovan C, Apweiler R (2009) QuickGO: a web-based tool for Gen Ontology searching. Bioinformatics 25(22):3045–3046
5. Xiang Z, Mungall C, Ruttenberg A, He Y (2011) Ontobee: a linked data server and browser for ontology terms. Proceedings of the 2nd international conference on biomedical ontologies (ICBO), Buffalo, NY, USA, 28–30 July 2011, pp 279–281
6. Mi H, Muruganujan A, Casagrande JT, Thomas PD (2013) Large-scale gene function analysis with the PANTHER classification system. Nat Protoc 8(8):1551–1566
7. Bauer S (2016) Gene-category analysis. In: Dessimoz C, Škunca N (eds) The gene ontology handbook. Methods in molecular biology, vol 1446. Humana Press. Chapter 13
8. Hastings J (2016) Primer on ontologies. In: Dessimoz C, Škunca N (eds) The gene ontology handbook. Methods in molecular biology, vol 1446. Humana Press. Chapter 1
9. Hastings J, de Matos P, Dekker A, Ennis M, Harsha B, Kale N, Muthukrishnan V, Owen G, Turner S, Williams M et al (2013) The ChEBI reference database and ontology for biologically relevant chemistry: enhancements for 2013. Nucleic Acids Res 41:D456–D463
10. Mungall C, Torniai C, Gkoutos G, Lewis S, Haendel M (2012) Uberon, an integrative multi-species anatomy ontology. Genome Biol 13:R5
11. Cooper L, Walls RL, Elser J, Gandolfo MA, Stevenson DW, Smith B, Preece J, Athreya B, Mungall CJ, Rensing S et al (2013) The Plant Ontology as a tool for comparative plant anatomy and genomic analyses. Plant Cell Physiol 54:e1
12. Berardini TZ, Khodiyar VK, Lovering RC, Talmud P (2010) The Gene Ontology in 2010: extensions and refinements. Nucleic Acids Res 38(Database Issue):D331–D335
13. Mungall CJ, Dietze H, Osumi-Sutherland D (2014) Use of OWL within the gene ontology. In Keet M, Tamma V (eds) Proceedings of the 11th international workshop on owl: experiences and directions (OWLED 2014), Riva del Garda, Italy, 17–18 October 2014, pp 25–36. doi:10.1101/010090
14. Dietze H, Berardini TZ, Foulger RE, Hill DP, Lomax J, Osumi-Sutherland D, Roncaglia P, Mungall CJ (2014) TermGenie – a web-application for pattern-based ontology class generation. J Biomed Semantics 5:48. doi:10.1186/2041-1480-5-48
15. Lovering RC (2016) How does the scientific community contribute to gene ontology? In: Dessimoz C, Škunca N (eds) The gene ontology handbook. Methods in molecular biology, vol 1446. Humana Press. Chapter 7
16. Gaudet P, Škunca N, Hu JC, Dessimoz C (2016) Primer on the gene ontology. In: Dessimoz C, Škunca N (eds) The gene ontology handbook. Methods in molecular biology, vol 1446. Humana Press. Chapter 3
17. Poux S, Gaudet P (2016) Best practices in manual annotation with the gene ontology. In: Dessimoz C, Škunca N (eds) The gene ontology handbook. Methods in molecular biology, vol 1446. Humana Press. Chapter 4
18. Škunca N, Roberts RJ, Steffen M (2016) Evaluating computational gene ontology annotations. In: Dessimoz C, Škunca N (eds) The gene ontology handbook. Methods in molecular biology, vol 1446. Humana Press. Chapter 8
19. Chibucos MC, Siegele DA, Hu JC, Giglio M (2016) The evidence and conclusion ontology (ECO): supporting GO annotations. In: Dessimoz C, Škunca N (eds) The gene ontology handbook. Methods in molecular biology, vol 1446. Humana Press. Chapter 18
20. Furnham N (2016) Complementary sources of protein functional information: the far side of GO. In: Dessimoz C, Škunca N (eds) The gene ontology handbook. Methods in molecular biology, vol 1446. Humana Press. Chapter 19

Advanced Gene Ontology Topics

17

The Evidence and Conclusion Ontology (ECO): Supporting GO Annotations

Marcus C. Chibucos, Deborah A. Siegele, James C. Hu, and Michelle Giglio

Abstract

The Evidence and Conclusion Ontology (ECO) is a community resource for describing the various types of evidence that are generated during the course of a scientific study and which are typically used to support assertions made by researchers. ECO describes multiple evidence types, including evidence resulting from experimental (i.e., wet lab) techniques, evidence arising from computational methods, statements made by authors (whether or not supported by evidence), and inferences drawn by researchers curating the literature. In addition to summarizing the evidence that supports a particular assertion, ECO also offers a means to document whether a computer or a human performed the process of making the annotation. Incorporating ECO into an annotation system makes it possible to leverage the structure of the ontology such that associated data can be grouped hierarchically, users can select data associated with particular evidence types, and quality control pipelines can be optimized. Today, over 30 resources, including the Gene Ontology, use the Evidence and Conclusion Ontology to represent both evidence and how annotations are made.

Key words Annotation, Biocuration, Conclusion, Confidence, Evidence, ECO, Experiment, Inference, Literature curation, Quality control

1 Describing Evidence in Scientific Investigations

1.1 Importance of Documenting Evidence

Investigations in the life sciences routinely produce data from diverse methodologies using a wide range of tools and techniques. Such data generated during the course of a research project contribute to the pool of evidence that ultimately leads a scientific researcher to make a particular inference or draw a given conclusion. Ultimately, one goal of a scientist is to publish the conclusions that are drawn from a given research project in the scientific literature. Such conclusions typically take the form of assertions, i.e., statements that are believed to be true, about some aspect of biology. The process of biocuration seeks to extract from the literature the **assertion** that summarizes the research finding *in addition to* any relevant **evidence** in support of the finding. Ideally, both of

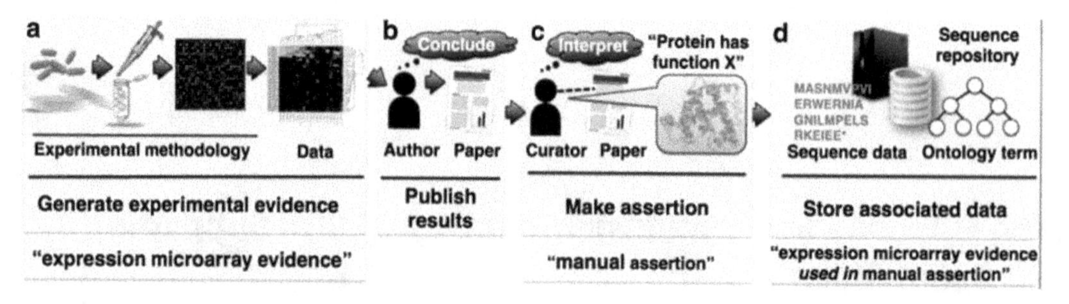

Fig. 1 Representing experimental methods and conclusions in a biological database. (**a**) An experiment is performed that generates data. (**b**) A researcher interprets methods and data, and draws conclusions that are published in a scientific journal and indexed in PubMed, for example. (**c**) A biocurator reads that paper, interprets the results presented therein, and makes an assertion. (**d**) The assertion is represented by associating an ontology term with the item being studied and stored along with other data, for example a protein sequence, at a biological database. (General summaries and related ECO classes are depicted along the bottom.)

these pieces of information will become integrated into a database in a structured way, so that they are readily accessible to the scientific community [1, 2] (Fig. 1).

Recording evidence is essential because: (1) knowing what methodologies were used is central to the scientific method and can impact one's evaluation of the data or results; (2) associating evidence with data maintained electronically allows for selective data queries and retrieval from even the largest of databases; and (3) a structured representation of evidence makes automated quality control possible, which is absolutely essential to managing the ever-increasing number and size of biological databases.

1.2 Multiple Types of Evidence and Ways of Associating Evidence with Assertions

Evidence can be associated with assertions in many ways. Manual curation is a common approach [3, 4], outlined in Fig. 1. However, text mining or other computational methods can also be used to extract biological assertions from the scientific literature [5, 6], and assertions can also be made directly via bioinformatic techniques [7], e.g. assigning of functional annotations as resulting from a functional genome annotation pipeline.

Numerous types of evidence form the bases for assertions that are made by researchers. Laboratory and field experiments are common sources of evidence, but computational (or *in silico*) analysis, whether executed by a person or an unsupervised machine, can also generate the evidence that is used to support assertions about biological function (Fig. 2). In addition, conclusions can be synthesized from investigator speculation or implied by known biology during the literature curation process. We can also consider *provenance*, a concept related to and sometimes conflated with evidence. A central goal of biological data repositories is to record in a structured fashion as much information as is known about the origins of a given accession. Yet sometimes an accession is imported from another database where the source for the annotation at that database is

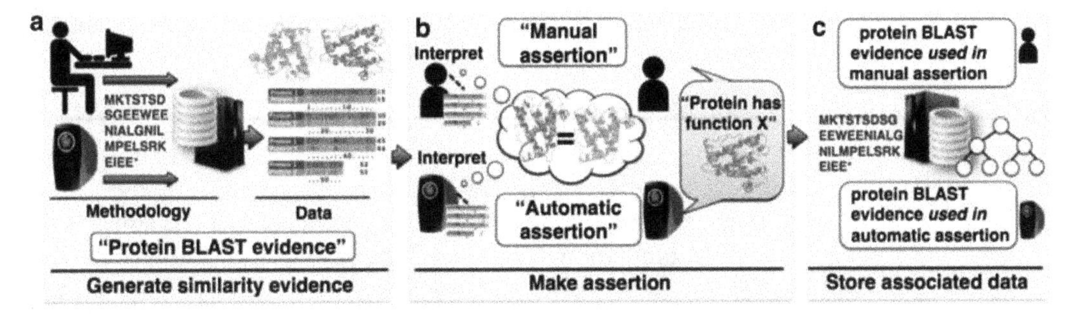

Fig. 2 Computational evidence and assertion. (**a**) A human or computer performs an analysis, for example comparing the sequence of a protein of unknown function to sequences at a database. A protein of known function is returned as a hit with corresponding alignment. (**b**) The alignment is analyzed and the protein sequences are deemed to share enough similarity to be considered homologs (related through common evolutionary descent). The query protein is assigned the same function as the database protein. (**c**) This information is stored at a sequence repository along with other data and metadata. (*Text in white boxes* depicts evidence and assertion methods used in this process.)

unclear. Even in this case it might be useful for the importing database to note the source of the statement/annotation along with a description of "imported information," indicating that nothing else is known about the evidence or provenance of that particular annotation. Thus there are numerous advantages to capturing scientific evidence and provenance, from describing specific methodologies to representing chains of custody.

2 The Evidence and Conclusion Ontology (ECO)

2.1 The Argument for an Ontology of Evidence

Due to the diversity of ways that exist to describe the multitude of scientific research methodologies, a means of representing evidence in a descriptive but structured way is required in order to maximize utility. The most efficient way to achieve this is to use an ontology, a controlled vocabulary where each term is well-defined and linked to other terms via defined relationships [8, 9]. In an ontological framework, evidence descriptions are represented not as free text, but rather as networked ontology classes where each child term is more specific (granular) than its parent [10]. High-level descriptions of types of evidence (such as "experimental evidence") are contained in more basal classes closest to the root class *evidence*. Increasingly specific terms that are grouped under the more general classes describe particular sub-types of evidence (such as "chromatography evidence"). The most specific terms, the so-called "leaf nodes" that contain no child terms, represent the most granular types of evidence generated during the course of a scientific investigation (for example "thin layer chromatography evidence"). **The Evidence and Conclusion Ontology (ECO)** (http://evidenceontology.org) was created to enable the structured description of experimental, computational, and other evidence types to support the assertions captured by scientific databases [11].

2.2 A Brief History of ECO

As described throughout this book, the Gene Ontology (GO) uses terms organized into controlled vocabularies, and the relationships among these terms, to capture functional information about gene products. The need to systematically document evidence while curating annotations was recognized from the inception of the GO [12] and a set of "evidence codes" was created for this purpose [13]. In time it was realized that a better-structured and more comprehensive way to represent evidence was required. Thus, the set of initially created GO codes, along with terms created by two model organism databases, FlyBase [14] and The *Arabidopsis* Information Resource [15], evolved into the first version of ECO, the "Evidence Code Ontology". Since then, the use of ECO by other resources has continued to grow and the ontology has shifted its focus beyond GO in order to become a generalized ontology for the capture of evidence information. The official name of ECO is now the "Evidence and Conclusion Ontology". ECO is presently being developed to define and broaden its scope, normalize its content, and enhance interoperability with related resources. The GO remains an active user and participant in developing ECO. It is anticipated that soon the three letter GO evidence codes to which so many are accustomed will be replaced by ECO term identifiers.

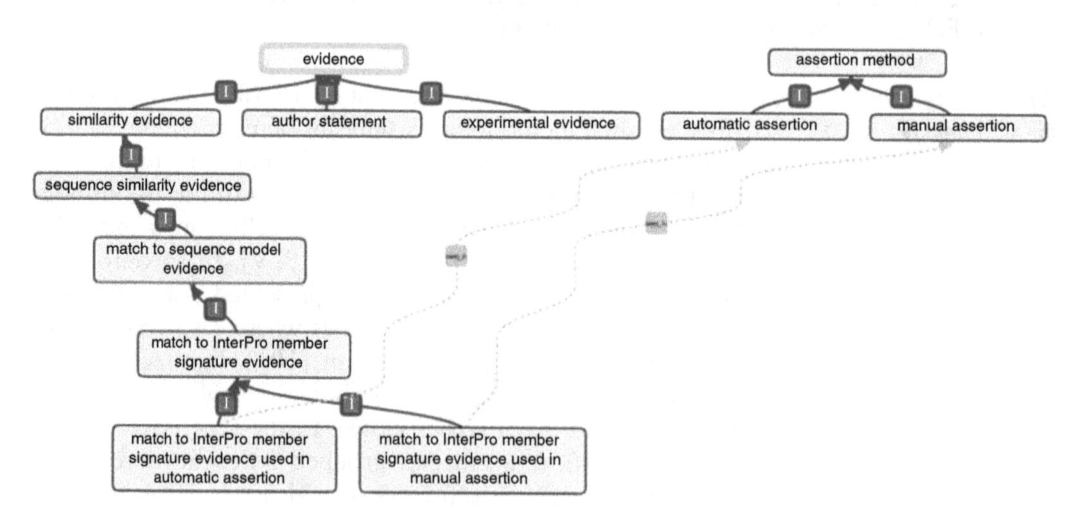

Fig. 3 Simplified representation of ECO, depicting general structure. ECO comprises two root classes along with their respective hierarchies, *evidence* (terms in *black*) and *assertion method* (terms in *pink*). A given type of evidence can be applied to (*used_in*; *dotted lines*) automatic assertion or manual assertion, which necessitated the creation of ECO leaf nodes that are *evidence* x *assertion method* cross products. For simplicity, most ECO classes are not displayed in the figure, including, for example, five of eight direct subclasses of *evidence* or three of four types of *similarity evidence* and so on

2.3 ECO Structure and Content

Evidence terms descend from the root class "evidence", which is defined as "a type of information that is used to support an assertion" (Fig. 3). Most evidence terms are either experimental or computational in nature, e.g., "chromatography evidence" or "sequence similarity evidence", respectively (Fig. 3). However, ECO also comprises other types of evidence, such as "curator inference" and "author statement".

In addition to describing evidence, ECO can also describe the means by which assertions are made, i.e., by a human or a machine. ECO calls this the "assertion method" and defines it as "a means by which a statement is made about an entity" (Figs. 1c and 2b). For example, whether a curator makes an annotation after reading about an experimental result in a scientific paper or after manually evaluating pairwise sequence alignment results, ECO can express that a manual curation method was used (3,8). Conversely, if an algorithm was used to assign a predicted function to a protein, ECO can express that an automated computational method was used. Thus "assertion method" forms a second root class with two branches: "manual assertion" and "automatic assertion" (Fig. 3).

The current version of ECO comprises 630 terms that describe "evidence", "assertion method", or "evidence x assertion method" cross products. Ontology architecture of ECO was recently described in Chibucos et al. [11].

2.3.1 Extending ECO Beyond GO

Recent development efforts of ECO have emphasized meeting the needs of a larger research community; *see* for example [11, 16], while still capturing the needed information for GO annotation, such as by adding comments and synonyms to a term. Many high-level ECO term definitions were written with explicit GO usage notes contained therein because ECO originated during early efforts of the GO. However, in order to increase overall usability of ECO by resources other than the GO, such verbiage has been removed, while retaining the essence of the term's meaning and applicability to GO. As ECO has been developed, more and more granular terms have been created to represent increasingly complex laboratory, computational, and even inferential techniques.

A discussion of ECO and GO would not be complete without mention of the GO evidence code IEA or "inferred from electronic annotation". IEA is used to connote that an annotation was assigned through automated computational means, e.g., transferring annotations from one protein to another. Because IEA describes how an annotation was *assigned*, rather than the specific type of supporting evidence, this term belongs as a subclass of "assertion method". As described above, "assertion method" has two child terms, "manual assertion" and "automatic assertion", with the latter being equivalent to IEA. Now it is possible to more accurately model evidence and the annotation process using ECO.

Aside from rewording definitions and creating a second root class, the biggest conceptual modification of ECO is reflected by removal of the prefix "inferred from" from every term name (see the GO codes for a sense of how ECO terms were previously labeled). This was done because ECO considers not just inferences made during the curation process, per se, but other aspects of evidence documentation, such as what research methodologies were performed.

3 Fundamentals of Evidence-Based GO Annotation

Creating an association between a GO term and a gene product is the fundamental essence of the GO annotation process. Documenting the evidence for any given GO annotation is a critical component of this annotation process, and an annotation would be incomplete without the requisite evidence. In fact, evidence capture by the GO requires both a "GO evidence code" that describes in detail the type of work or analysis that was performed in support of the annotation, as well as a citation for the reference from which the evidence was derived. Curators go to great lengths to understand and properly apply the correct "evidence code" to a given annotation, and an online guide exists to explain the often-subtle distinctions between multiple related evidence types (http://geneontology.org/page/guide-go-evidence-codes) [4, 13].

The GO gene association file (GAF) format contains required columns for both evidence code and reference. Each GO evidence code maps directly to an ECO term. ECO maintains database cross references to the GO codes for easy mapping between systems. GO codes therefore represent a subset of the Evidence and Conclusion Ontology. Since independent development of ECO was undertaken, a number of new GO evidence codes have been created, e.g., IBA, IBD, IKR, IRD. Equivalent terms have been instantiated in ECO (Fig. 4a), which will continue to develop such terms for the GO.

3.1 ECO Terms Versus GO Codes

Although GO evidence codes are useful in themselves because they represent detailed descriptions of evidence types, they are maintained as a controlled vocabulary with a shallow hierarchical structure that lacks the advantages of a formal ontology like ECO. Further, the full set of terms within ECO provides the ability to capture more breadth and depth of evidence information than the GO evidence codes do. Additionally, as the field of biocuration evolves and the kinds of evidence being curated from the literature continue to grow both more detailed and nuanced, the number of two- and three-letter acronyms (e.g., IEA, IMP, EXP, and ISS) available for new terms will hit an upper limit (there are only 676 possibilities using all 26 two-letter combinations, as the first letter of the three-letter GO codes often stands for "inferred"). In fact, ECO developers have already received requests from different users to develop new, but unrelated, terms that had the same suggested three-letter acronyms. For all of these reasons, there are discussions underway about transitioning GO evidence storage to use ECO terms rather than GO evidence codes. Such a shift would combine the

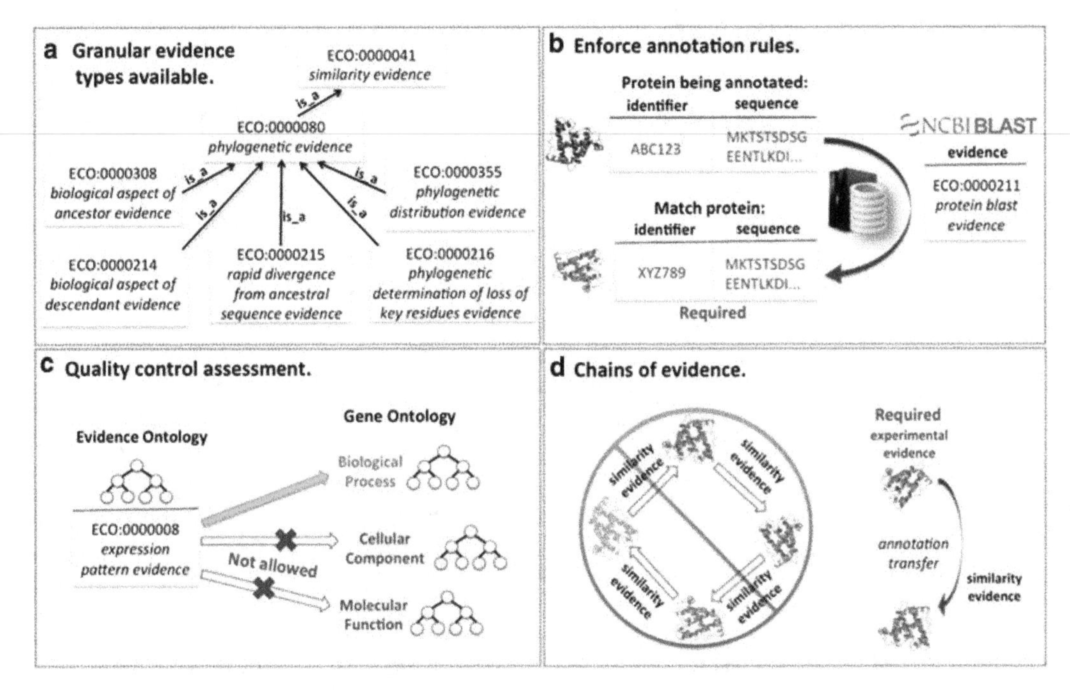

Fig. 4 Applications of ECO to GO. (**a**) ECO evidence classes are hierarchical such that broader classes parent more granular ones; depicted here are evidence types that support a phylogenetic tree-based approach for generating manually reviewed, homology-based annotations. (**b**) When a protein is annotated based on sequence similarity to another annotated protein, the identity of that protein must be recorded in the annotation file along with the evidence. (**c**) Quality control assessment: Expression pattern evidence is only allowable for annotations to the GO Biological Process ontology. (**d**) Evidence is used to prevent circular annotations based solely on computational predictions. Chains of evidence are computationally evaluated to ensure that inferential annotations are linked to experimental evidence

advantages of both systems and would still provide a mechanism for filtering evidence annotations by the previous codes if desired. If ECO terms were to be fully adopted by GO, the GAF format would change to require "ECO term" instead of "evidence code." Since most GO evidence codes have a one-to-one mapping to ECO terms (while the remainder, i.e., IEA, IGC, ISS, map, in conjunction with various GO standard references [http://purl.obolibrary.org/obo/eco/gaf-eco-mapping.txt], to specific ECO terms), GO data depositors could use a straightforward replacement based on the mappings. Other resources outside of GO have modeled their annotation capture systems on the GAF format. For example, the Ontology of Microbial Phenotypes [17] uses a modified version of the GO GAF, but employs ECO terms instead of GO evidence codes. The full use of ECO terms by the GO would enhance the integration of data derived from such diverse sources.

4 Benefits of ECO and Applications for the GO

There are currently over 365 million annotations in the GO repository linked to an evidence term, and these can be queried and maintained better with the help of an ontology by leveraging its hierarchical structure. One of the most direct applications for using an ontology of evidence is *selective data query*, i.e., to query a database for records associated with a particular evidence type. For example, searching for "thin layer chromatography evidence" (at present a leaf term with no subclasses) would return only the records associated with that evidence type and no others. But *grouping annotations* is also possible with this approach. A query for "chromatography evidence" will return data associated not only with "chromatography evidence" but also its more specific subtypes including "thin layer chromatography evidence" and "high performance liquid chromatography evidence".

But there are further benefits to be derived from an ontology of evidence beyond simple structured queries (Fig. 4). For example:

1. To amplify the benefits of experimental knowledge that curators capture, the GO Consortium is using a phylogenetic tree-based approach to generate manually reviewed, homology-based annotations for a range of species [18]. This phylogenetic annotation methodology necessitated a new set of evidence terms to capture the inference process (Fig. 4a). Currently over 150,000 annotations are associated with these new terms and the number continues to grow.
2. The GO curatorial process uses evidence to support computable rules about the kinds of information that must be associated with different evidence types. For example, one rule states that annotation of a protein based on alignment with another protein requires that the identity of the matching protein be captured, along with the evidence type "protein alignment evidence" (Fig. 4b). If such an evidence type were missing, this would flag the annotation for review.
3. The GO uses evidence as a quality control mechanism for annotation consistency. For example, expression pattern evidence is restricted to annotations for terms from the "biological process" ontology. Annotations to terms from either of the other two GO ontologies ("molecular function" or "cellular component") would be flagged as suspect (Fig. 4c).
4. Evidence is used to prevent circular annotations based solely on computational predictions (Fig. 4d). Chains of evidence are computationally evaluated to ensure that inferential annotations are linked to experimental evidence. For example, annotations supported by "sequence alignment evidence" require the inclusion of a database identifier for the matching gene

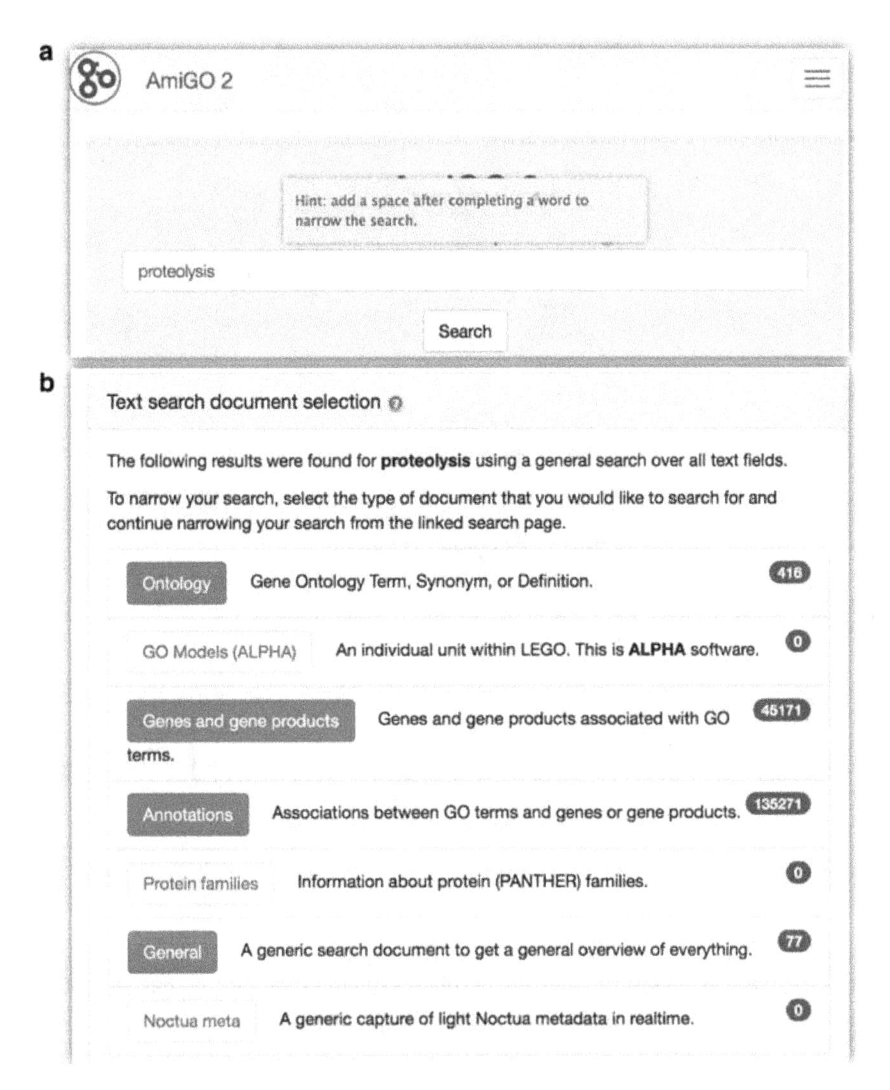

Fig. 5 AmiGO 2 query and results. (**a**) User has typed "proteolysis" into the search box. (**b**) Number of hits (*right gray box*) shown for each document category (*blue boxed text*). Clicking on "Annotations" will open a new page with more detailed results

product that is itself linked to an annotation supported by experimental evidence.

Yet another application of ECO for the GO has been realized in the UniProt-Gene Ontology Annotation (UniProt-GOA) project. Arguably, UniProt is the most comprehensive and best-curated protein database available to the research community. ECO terms have replaced the original UniProtKB [19] evidence types and are available in UniProtKB XML [11]. Novel ways of mapping and extending ontologies have been discussed with ECO and the GO Consortium to ensure appropriate development for UniProtKB annotation. The

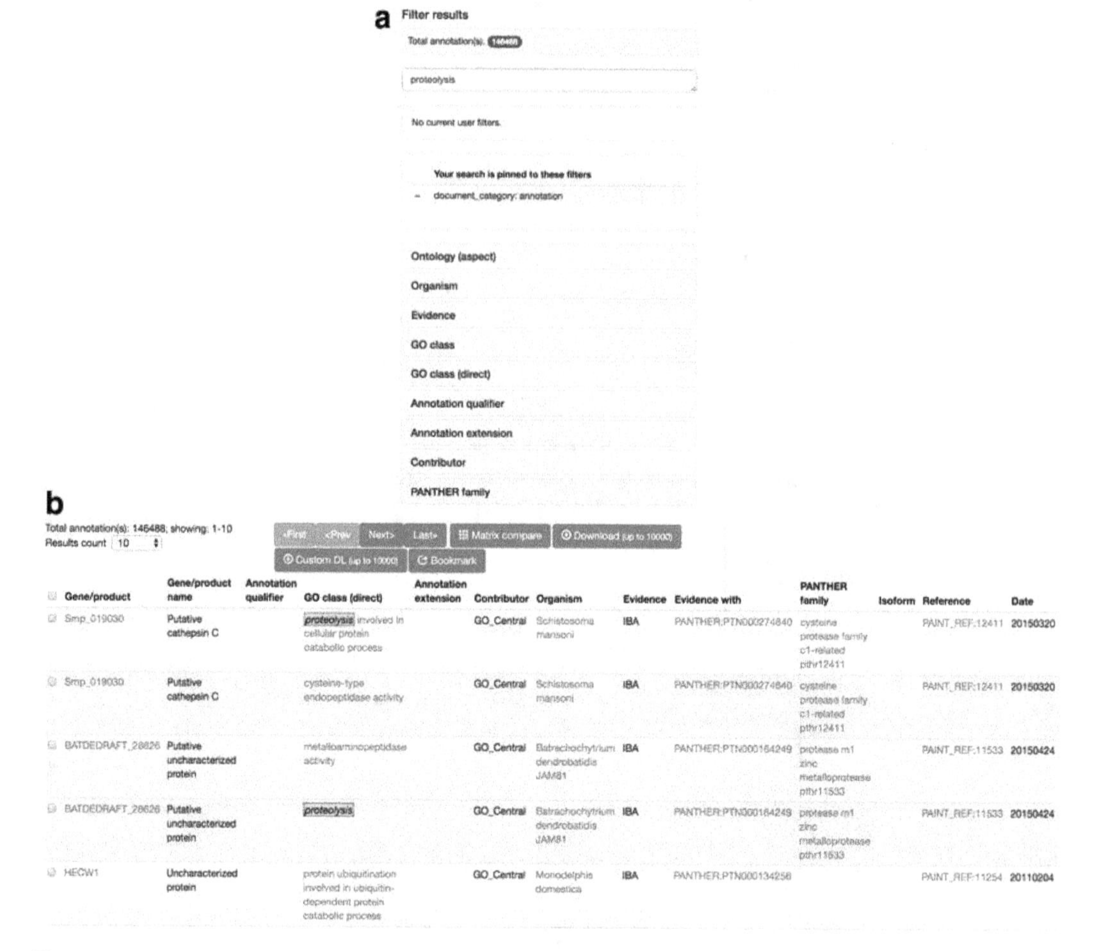

Fig. 6 Annotation hits to a query search. (**a**) To the left of the search results, the user has an opportunity to click on filters. (**b**) To the right, each annotation row is shown for a given protein

UniProt-GOA project provides >169 million manual and electronic evidence-based associations between GO terms and 26.5 million UniProtKB proteins covering >411,000 taxa [20]. Of these, manual annotation provides 1.4 million annotations to ~260,000 proteins. Since 2010, UniProt-GOA has supplied GO annotations in a Gene Product Association Data (GPAD) file format, which allows inclusion of ECO terms. Because ECO terms are cross referenced to corresponding GO codes, even if evidence for annotations was supplied to UniProt as GO codes, the GPAD file will display the appropriate equivalent ECO term. Thus, UniProt annotations can be grouped by leveraging the structure of ECO.

4.1 Exercise

Once the reader has gained a basic understanding of ECO and its connection to GO, we can perform the following simple exercise

Evidence

Count	Evidence type
(80346)	similarity evidence used in manual assertion
(80049)	biological aspect of ancestor evidence used in manual assertion
(41834)	evidence used in automatic assertion
(8086)	experimental evidence
(7587)	sequence similarity evidence
(4058)	direct assay evidence used in manual assertion
(3906)	sequence orthology evidence
(3574)	mutant phenotype evidence used in manual assertion
(2743)	traceable author statement used in manual assertion
(1057)	match to sequence model evidence
(399)	non-traceable author statement used in manual assertion
(351)	computational combinatorial evidence used in manual assertion
(279)	genetic interaction evidence used in manual assertion
(177)	inference from background scientific knowledge
(168)	sequence alignment evidence used in manual assertion
(129)	phylogenetic determination of loss of key residues evidence used in manual assertion
(71)	physical interaction evidence used in manual assertion
(14)	expression pattern evidence
(1)	genomic context evidence used in manual assertion

Fig. 7 Selected ECO terms in use by the GO Consortium that are related to the present query. The number of annotations supported by a given evidence type is shown in *parentheses*

that displays a faceted query using ECO in AmiGO 2 (http://amigo2.geneontology.org/amigo).

User types "proteolysis" into the query box (Fig. 5a) and sees a number of hits returned (Fig. 5b). Next, after clicking on "Annotations" in the blue rectangle, the user sees all the annotation-related terms that had hits to "proteolysis" (Fig. 6a, b), split into two parts here for easier viewing. Clicking on "Evidence" in the filter box (Fig. 6a) will expand it to display all constituent evidence types (Fig. 7). Clicking on

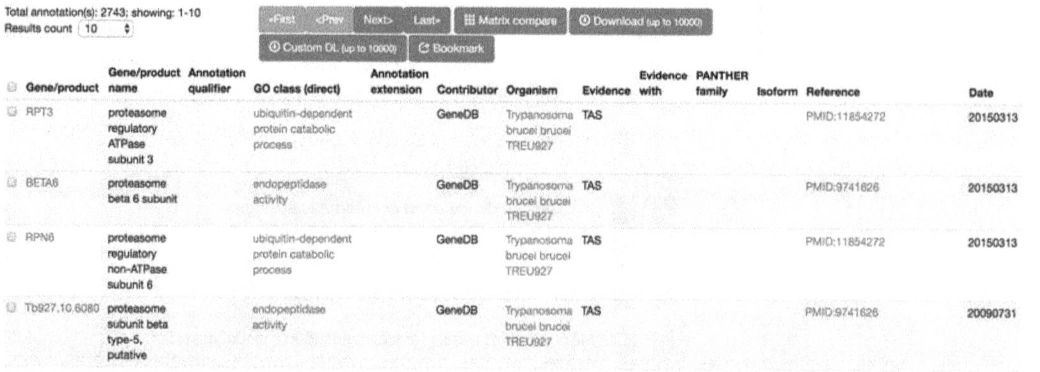

Total annotation(s): 2743; showing: 1-10
Results count 10

«First <Prev Next> Last» Matrix compare Download (up to 10000)
Custom DL (up to 10000) Bookmark

Gene/product	Gene/product name	Annotation qualifier	GO class (direct)	Annotation extension	Contributor	Organism	Evidence	Evidence with	PANTHER family	Isoform	Reference	Date
RPT3	proteasome regulatory ATPase subunit 3		ubiquitin-dependent protein catabolic process		GeneDB	Trypanosoma brucei brucei TREU927	TAS				PMID:11854272	20150313
BETA6	proteasome beta 6 subunit		endopeptidase activity		GeneDB	Trypanosoma brucei brucei TREU927	TAS				PMID:9741626	20150313
RPN6	proteasome regulatory non-ATPase subunit 6		ubiquitin-dependent protein catabolic process		GeneDB	Trypanosoma brucei brucei TREU927	TAS				PMID:11854272	20150313
Tb927.10.6080	proteasome subunit beta type-5, putative		endopeptidase activity		GeneDB	Trypanosoma brucei brucei TREU927	TAS				PMID:9741626	20090731

Fig. 8 Filtering on evidence. After filtering on "traceable author statement used in manual assertion", only annotations supported by that evidence type are displayed, shown as "TAS" in the "Evidence" column. Number of annotations associated with that evidence type is shown at the *top left*

"traceable author statement used in manual assertion" will open a subset of the results that match that more restrictive filter (Fig. 8). The evidence filter box now says "Nothing to filter" (Fig. 9).

5 The Future of ECO

What else can an ontology of evidence do? One aspect of active exploration for ECO is the evaluation of confidence or quality of evidence. Work has begun [21] to develop a mechanism to incorporate quality information into ECO or, as needed, to create a standalone system. It might one day be possible to use ECO to describe the *quality* of the evidence supporting an annotation in addition to the *type* of evidence that supports the annotation.

In summary, the Evidence and Conclusion Ontology can be used to support faceted queries of data, to establish computable rules about required types of evidence, as a quality control check for annotation consistency, and as a mechanism to prevent circular annotations rooted only in computational predictions. GO is already benefitting from these applications of ECO, and the future promises both additional new applications of ECO as well as advancements to current ones.

Acknowledgements

This material is based upon work supported by the National Science Foundation under Award Number 1458400 and National

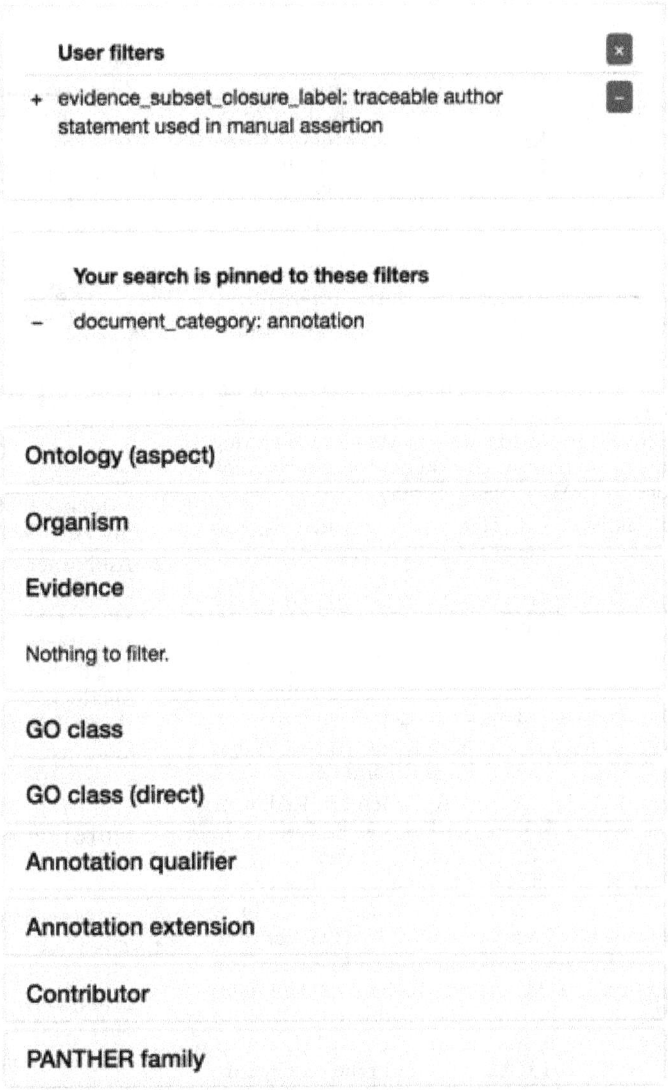

User filters

+ evidence_subset_closure_label: traceable author statement used in manual assertion

Your search is pinned to these filters

- document_category: annotation

Ontology (aspect)

Organism

Evidence

Nothing to filter.

GO class

GO class (direct)

Annotation qualifier

Annotation extension

Contributor

PANTHER family

Fig. 9 Filter box fully selected. Increasingly granular evidence filters have been applied until there is nothing left to filter

Institutes of Health/National Institute of General Medical Sciences under Grant Number 2R01 GM089636. Open Access charges were funded by the University College London Library, the Swiss Institute of Bioinformatics, the Agassiz Foundation, and the Foundation for the University of Lausanne.

References

1. Gaudet P, Arighi C, Bastian F, Bateman A, Blake JA, Cherry MJ, D'Eustachio P, Finn R, Giglio M, Hirschman L, Kania R, Klimke W, Martin MJ, Karsch-Mizrachi I, Munoz-Torres M, Natale D, O'Donovan C, Ouellette F, Pruitt KD, Robinson-Rechavi M, Sansone SA, Schofield P, Sutton G, Van Auken K, Vasudevan S, Wu C, Young J, Mazumder R (2012) Recent advances in biocuration: meeting report from the Fifth International Biocuration Conference. Database:bas036. doi:10.1093/database/bas036
2. Burge S, Attwood TK, Bateman A, Berardini TZ, Cherry M, O'Donovan C, Xenarios L, Gaudet P (2012) Biocurators and biocuration: surveying the 21st century challenges. Database:bar059. doi:10.1093/database/bar059
3. Balakrishnan R, Harris MA, Huntley R, Van Auken K, Cherry JM (2013) A guide to best practices for Gene Ontology (GO) manual annotation. Database:bat054. doi:10.1093/database/bat054
4. Poux S, Gaudet P (2016) Best practices in manual annotation with the gene ontology. In: Dessimoz C, Škunca N (eds) The gene ontology handbook. Methods in molecular biology, vol 1446. Humana Press. Chapter 4
5. Arighi CN, Carterette B, Cohen KB, Krallinger M, Wilbur WJ, Fey P, Dodson R, Cooper L, Van Slyke CE, Dahdul W, Mabee P, Li D, Harris B, Gillespie M, Jimenez S, Roberts P, Matthews L, Becker K, Drabkin H, Bello S, Licata L, Chatr-aryamontri A, Schaeffer ML, Park J, Haendel M, Van Auken K, Li Y, Chan J, Muller HM, Cui H, Balhoff JP, Chi-Yang Wu J, Lu Z, Wei CH, Tudor CO, Raja K, Subramani S, Natarajan J, Cejuela JM, Dubey P, Wu C (2013) An overview of the BioCreative 2012 Workshop Track III: interactive text mining task. Database:bas056. doi:10.1093/database/bas056
6. Altman RB, Bergman CM, Blake J, Blaschke C, Cohen A, Gannon F, Grivell L, Hahn U, Hersh W, Hirschman L, Jensen LJ, Krallinger M, Mons B, O'Donoghue SI, Peitsch MC, Rebholz-Schuhmann D, Shatkay H, Valencia A (2008) Text mining for biology--the way forward: opinions from leading scientists. Genome Biol 9(Suppl 2):S7. doi:10.1186/gb-2008-9-s2-s7
7. Cozzetto D, Jones DT (2016) Computational methods for annotation transfers from sequence. In: Dessimoz C, Škunca N (eds) The gene ontology handbook. Methods in molecular biology, vol 1446. Humana Press. Chapter 5
8. Smith B (2003) Ontology. In: Floridi L (ed) Blackwell guide to the philosophy of computing and information. Blackwell, Oxford, pp 155–166
9. Smith B (2008) Ontology (Science). In: Eschenbach C, Grüninger M (eds) Formal ontology in information systems. Ios Press, Amsterdam, pp 21–35
10. Hastings J (2016) Primer on ontologies. In: Dessimoz C, Škunca N (eds) The gene ontology handbook. Methods in molecular biology, vol 1446. Humana Press. Chapter 1
11. Chibucos MC, Mungall CJ, Balakrishnan R, Christie KR, Huntley RP, White O, Blake JA, Lewis SE, Giglio M (2014) Standardized description of scientific evidence using the Evidence Ontology (ECO). Database:bau075. doi:10.1093/database/bau075
12. Ashburner M, Ball CA, Blake JA, Botstein D, Butler H, Cherry JM, Davis AP, Dolinski K, Dwight SS, Eppig JT, Harris MA, Hill DP, Issel-Tarver L, Kasarskis A, Lewis S, Matese JC, Richardson JE, Ringwald M, Rubin GM, Sherlock G (2000) Gene ontology: tool for the unification of biology. The Gene Ontology Consortium. Nat Genet 25(1):25–29. doi:10.1038/75556
13. Gaudet P, Škunca N, Hu JC, Dessimoz C (2016) Primer on the gene ontology. In: Dessimoz C, Škunca N (eds) The gene ontology handbook. Methods in molecular biology, vol 1446. Humana Press. Chapter 3
14. The FlyBase Consortium (2002) The FlyBase database of the *Drosophila* genome projects and community literature. Nucleic Acids Res 30(1):106–108
15. Huala E, Dickerman AW, Garcia-Hernandez M, Weems D, Reiser L, LaFond F, Hanley D, Kiphart D, Zhuang M, Huang W, Mueller LA, Bhattacharyya D, Bhaya D, Sobral BW, Beavis W, Meinke DW, Town CD, Somerville C, Rhee SY (2001) The *Arabidopsis* Information Resource (TAIR): a comprehensive database and web-based information retrieval, analysis, and visualization system for a model plant. Nucleic Acids Res 29(1):102–105
16. Kilic S, White ER, Sagitova DM, Cornish JP, Erill I (2014) CollecTF: a database of experimentally validated transcription factor-binding sites in Bacteria. Nucleic Acids Res 42(Database issue):D156–D160. doi:10.1093/nar/gkt1123
17. Chibucos MC, Zweifel AE, Herrera JC, Meza W, Eslamfam S, Uetz P, Siegele DA, Hu JC, Giglio MG (2014) An ontology for microbial

phenotypes. BMC Microbiol 14(1):294. doi:10.1186/s12866-014-0294-3

18. Reference Genome Group of the Gene Ontology Consortium (2009) The Gene Ontology's Reference Genome Project: a unified framework for functional annotation across species. PLoS Comput Biol 5(7):e1000431. doi:10.1371/journal.pcbi.1000431
19. UniProt Consortium (2014) Activities at the Universal Protein Resource (UniProt). Nucleic Acids Res 42(Database issue):D191–D198. doi:10.1093/nar/gkt1140
20. Dimmer EC, Huntley RP, Alam-Faruque Y, Sawford T, O'Donovan C, Martin MJ, Bely B, Browne P, Mun Chan W, Eberhardt R, Gardner M, Laiho K, Legge D, Magrane M, Pichler K, Poggioli D, Sehra H, Auchincloss A, Axelsen K, Blatter MC, Boutet E, Braconi-Quintaje S, Breuza L, Bridge A, Coudert E, Estreicher A, Famiglietti L, Ferro-Rojas S, Feuermann M, Gos A, Gruaz-Gumowski N, Hinz U, Hulo C, James J, Jimenez S, Jungo F, Keller G, Lemercier P, Lieberherr D, Masson P, Moinat M, Pedruzzi I, Poux S, Rivoire C, Roechert B, Schneider M, Stutz A, Sundaram S, Tognolli M, Bougueleret L, Argoud-Puy G, Cusin I, Duek-Roggli P, Xenarios I, Apweiler R (2012) The UniProt-GO Annotation database in 2011. Nucleic Acids Res 40(Database issue):D565–D570. doi:10.1093/nar/gkr1048
21. Bastian FB, Chibucos MC, Gaudet P, Giglio M, Holliday GL, Huang H, Lewis SE, Niknejad A, Orchard S, Poux S, Skunca N, Robinson-Rechavi M (2015) The Confidence Information Ontology: a step towards a standard for asserting confidence in annotations. Database:bav043. doi:10.1093/database/bav043

18

Annotation Extensions

Rachael P. Huntley and Ruth C. Lovering

Abstract

The specificity of knowledge that Gene Ontology (GO) annotations currently can represent is still restricted by the legacy format of the GO annotation file, a format intentionally designed for simplicity to keep the barriers to entry low and thus encourage initial adoption. Historically, the information that could be captured in a GO annotation was simply the role or location of a gene product, although genetically interacting or binding partners could be specified. While there was no mechanism within the original GO annotation format for capturing additional information about the context of a GO term, such as the target gene of an activity or the location of a molecular function, the long-term vision for the GO Consortium was to provide greater expressivity in its annotations to capture physiologically relevant information.

Thus, as a step forwards, the GO Consortium has introduced a new field into the annotation format, ***annotation extensions***, which can be used to capture valuable contextual detail. This provides experimentally verified links between gene products and other physiological information that is crucial for accurate analysis of pathway and network data. This chapter will provide a simple overview of annotation extensions, illustrated with examples of their usage, and explain why they are useful for scientists and bioinformaticians alike.

Key words Gene Ontology, Annotation, Biocuration, Context, Pathway, Network, Analysis, Annotation extension

1 Introduction

Functional annotation of gene products using the GO has gone far in simplifying the task of finding functional roles of both individual and groups of gene products. It has enabled a multitude of analyses that were previously not possible. For example, GO annotations are invaluable for analyzing a list of genes that are identified as differentially expressed in a microarray experiment using one of the many freely available functional enrichment programs [1, 2] (*see* also Chap. 13 [3]).

The original simplistic GO annotation pairs a gene product with a GO term (one of ***biological process***, ***molecular function*** or ***cellular component***). Because these pair-wise associations are treated independently, vast amounts of correlated functional data are omitted from the basic GO annotation and therefore inaccessible to network and pathway analyses. This contextual information is essential for understanding the physiological roles of gene products. Without contextual information bioinformatics analyses cannot identify gene products that perform a role only under certain conditions or in the presence of specific factors and therefore will present an incomplete view of the available data [4]. Specific gene products will often have different biological roles in different cells or tissues as these roles will be dependent on the available interacting partners; already tissue-specific network analyses are able to demonstrate the importance of the cellular environment. For example, Greene et al. [5] analyzed the GO and pathway annotations of the available interaction partners of the transcription factor, LEF1, in different tissue types. They demonstrated that LEF1 was significantly associated with biological processes that were relevant to each tissue type. For instance, in blood vessels the LEF1 interacting partners were associated with angiogenesis, whereas in hypothalamus they are associated with hypothalamus development.

Here we describe an incremental extension of the GO annotation format to allow more detailed statements about gene product function, which will benefit all types of functional analyses [5].

2 Extending the Core GO Annotation Model

In practical terms, the newly introduced *annotation extensions* field enables curators to provide appropriate experimentally evidenced contextual information for manually curated annotations (extant software pipelines for electronically inferred annotations (IEA) do not yet support population of this field).

Generating a comprehensive annotation, one that includes its context, involves refining the core pair-wise association with additional relationships to other ontology classes [5]. This dynamic approach is logically equivalent to creating a new term for the subtype in the ontology, but offers advantages in terms of both flexibility and efficiency.

In essence this approach allows curators to dynamically create "virtual" terms. It enables curators to combine all of the specific terms needed to fully describe a gene product in a way that can be reproducibly, computationally interpreted. For example, "core RNA polymerase binding transcription factor in hypothalamus", associates a gene product with that activity occurring in that specific location. From the computer logic perspective this effectively has created a subclass of "core RNA polymerase binding transcription factor activity" (GO:0000990). The flexibility of expression thus supports the virtual creation of complex, compound child terms on an as-needed basis. Additionally, this approach to virtual term creation is immediate. Because the parent can be automatically inferred from the primary term of the association, and because the additional relationships to other terms provide the refinements needed to create a more specific term, the result is that the previously independent processes of annotating gene products and creating ontology terms are now fully integrated. The use of annotation extensions means that curators can immediately make the biological statement required without having to return to the annotation to update it only after the term is available in the ontology, thus making the overall process more efficient. As these virtual terms are not consequently added to the ontology—although they could be if required—the extended annotations can be "folded" to create the logical equivalent of a GO term [5]. The GO Consortium (GOC) is in the process of incorporating these inferred annotations into the files it provides and so this contextual information will be included by default for use by anyone, or any analysis tool, that utilizes the annotation files.

3 Annotation Extension Format

Annotation extensions refine the GO term used in the basic annotation by adding one or more relational expressions (extensions). Each extension is written as *Relation(Entity)*, where *Relation* is a label describing the relationship between the GO term and the entity, and *Entity* is an identifier for a database object or ontology term, for example *part_of(GO:0005634)*, where GO:0005634 is the Gene Ontology identifier for "nucleus".

Relations can be one of two types: "molecular relations" that are used with entities such as a gene, gene product, complex, or chemical and "contextual relations" that are used with entities such as a cell type, anatomy term, developmental stage, or a GO term.

In order to clearly define the semantics of the extensions, rules have been implemented defining what types of entity identifiers may be used with each relation. Generally, curators may only use contextual relations (e.g., where and when) with terms from the Cell Type Ontology (CL) [6], Uber Anatomy Ontology (Uberon) [7], Plant Ontology (PO) [8], nematode life stages (WBls) [9] and certain GO terms, and molecular target relations may only apply to a physical entity such as a gene product (e.g., UniProtKB [10] or PomBase [11]), a macromolecular complex (e.g., Intact Complex Portal [12]), or a chemical using a ChEBI [13] identifier. Curation tools can incorporate these rules to prevent invalid annotations from being created. Table 1 shows the most commonly used relations with examples of their usage.

Table 1
Most commonly used relationships for annotation extension statements and examples of their usage

Contextual relationships	Example (gene product; primary GO term; annotation extension)
part_of	*C. elegans* psf-1; nucleus; part_of(WBbt:0006804 *body wall muscle cell*)
occurs_in	Mouse opsin-4; G-protein coupled photoreceptor activity; occurs_in(CL:0000740 *retinal ganglion cell*)
happens_during	*S. pombe* wis4; stress-activated MAPK cascade; happens_during(GO:0071470 *cellular response to osmotic stress*)
Molecular relationships	**Example (gene product; primary GO term; annotation extension)**
has_regulation_target	Human suppressor of fused homolog SUFU; negative regulation of transcription factor import into nucleus; has_regulation_target(UniProtKB:P08151 *zinc finger protein GLI1*)
has_input	*S. pombe* rlf2: protein localization to nucleus; has_input(PomBase:SPAC26H5.0 *pcf2*)
has_direct_input	Human WNK4; chloride channel inhibitor activity; has_direct_input(UniProtKB:Q7LBE3 *Solute carrier family 26 member 9*)

Molecular relations take an entity such as a gene, gene product, complex, or chemical as an argument; contextual relations take an entity such as a cell type, anatomy term, development stage, or a GO term as an argument. Entity names in italics are shown for clarity and are not part of the annotation extension format.

4 Improved Expressiveness of GO Annotations: Examples

4.1 Targets of an Enzyme

One means of adding value to a GO annotation, using annotation extensions, is by specifying the molecular target of an enzyme activity. The inability to add effector–target relationships has been a major limitation of the core GO annotation model, with this addition we can now begin to provide directional information that can be used for network and pathway analyses. Take as an example the annotation of human mitogen-activated protein kinase-activated protein kinase 2 (MAPKAP-K2), which was shown to phosphorylate the CapZ-interacting protein (CapZIP) [14]. A basic GO annotation would describe MAPKAP-K2 as a protein serine/threonine kinase:

Gene product:	UniProtKB:P49137 (*human MAPKAP-K2*)
GO term:	GO:0004674 (*protein serine/threonine kinase activity*)

Using an annotation extension, a curator can add more detail as follows:

Gene product:	UniProtKB:P49137 (*human MAPKAP-K2*)
GO term:	GO:0004674 (*protein serine/threonine kinase activity*)
Extension:	has_direct_input(UniProtKB:Q6JBY9) (*human CapZIP*)

N.B. phrases in italics are not part of the syntax but are added for better interpretation by the reader.

The extended GO annotation describes MAPKAP-K2 as a protein serine/threonine kinase that can phosphorylate CapZIP. This is vital information that can be utilized for linking together processes and pathways that MAPKAP-K2 and CapZIP, and any further targets of these proteins, are involved in. The rules of usage for has_direct_input are that the primary GO term used should be a Biological Process or Molecular Function and in this example the term used is a Molecular Function, additionally the entity used in the extension should be a gene product, macromolecular complex, or chemical and in this example it is a gene product, i.e., a protein. Note that has_direct_input was used here instead of has_input because there was evidence in the paper that MAPKAP-K2 acted directly on the substrate CapZIP, if there was a possibility of an intermediate molecule in this reaction, has_input would have been used.

4.2 Anatomical Location of a Gene Product's Function

An annotation can be extended to specify the locational context in which a gene product performs its roles. It is important to note that we intend only to capture those locations that are physiologically relevant to the organism and not the experimental detail in which the observation was made.

The rat protein dihydrofolate reductase (Dhfr) was shown to reduce dihydrofolic acid to tetrahydrofolic acid in rat neurons [15]. From this evidence a basic GO annotation could be made as follows:

Gene product:	UniProtKB:Q920D2 (*rat Dhfr*)
GO term:	GO:0004146 (*dihydrofolate reductase activity*)

By extending the annotation the curator can also specify in which cell type this activity occurs:

Gene product:	UniProtKB:Q920D2 (*rat Dhfr*)
GO term:	GO:0004146 (*dihydrofolate reductase activity*)
Extension:	occurs_in(CL:0000540) (*neuron*)

This annotation now provides the physiologically relevant information that Dhfr is active in neurons. The rules for occurs_in are that the primary GO term used must be a Biological Process or Molecular Function (in this example it is a Molecular Function); additionally the entity in the extension must be a cell type, anatomical feature, or GO Cellular Component (in this example it is an identifier from the Cell Type Ontology).

4.3 Timing-Specific Location of a Gene Product

A gene product's annotation may be made more specific by including the appropriate developmental stage. An example is the location of the *C. elegans* PAXT-1 protein, which is located in the nucleus during the embryo stage [16]. Using the basic GO annotation format, a curator might indicate that PAXT-1 is located in the nucleus:

Gene product:	UniProtKB:Q21738 (*C. elegans PAXT-1*)
GO term:	GO:0005634 (*nucleus*)

By extending the annotation the curator can also specify when this localization occurs:

Gene product:	UniProtKB:Q21738 (*C. elegans PAXT-1*)
GO term:	GO:0005634 (*nucleus*)
Extension:	exists_during(WBls:0000003) (*embryo*)

This annotation means that PAXT-1 is located in the nucleus during the *C. elegans* embryo stage. The rules for exists_during are that the primary GO term used should be a Cellular Component and the entity in the extension should be a developmental stage or a GO Biological Process, in this case the entity is from the *C. elegans* life stage ontology.

4.4 Multiple Relational Expressions

If several contextual statements can be made for the gene product, it is possible to combine relational expressions to make even more complex statements. Relational expressions can be separated by commas "," (meaning AND) or by pipes, "|" (meaning OR), depending on whether the conditions in the statement are co-occurring (AND) or independent (OR).

The human microRNA miR-145 provides an example of the application of multiple annotation extensions. MiR-145 was shown to directly bind and silence the POU5F1 transcription factor, among others, causing inhibition of embryonic stem cell division [17]. This evidence could therefore be represented by two basic GO annotations as follows:

Gene product:	RNACentral:URS0000527F89_9606 (*human miR-145*)
GO term:	GO:1903231 (*mRNA binding involved in posttranscriptional gene silencing*)
Gene product:	RNACentral:URS0000527F89_9606 (*human miR-145*)
GO term:	GO:1904676 (*negative regulation of somatic stem cell division*)

Using relational expressions, separated by commas, we can make one extended annotation as follows:

Gene product:	RNACentral:URS0000527F89_9606 (*human miR-145*)
GO term:	GO:1903231 (*mRNA binding involved in posttranscriptional gene silencing*)
Extension:	has_direct_input(Ensembl:ENSG00000204531), occurs_in(CL:0002322), part_of(GO:1904676) (*human POU5F1, embryonic stem cell, negative regulation of somatic stem cell division*)

The extended annotation signifies that miR-145 directly binds and silences POU5F1 mRNA expression as part of the inhibition of somatic stem cell division of embryonic stem cells. Again, this contextual information will be essential information when analyzing the physiological relevance of the role of a gene product in a pathway.

Although the use of a pipe (|) to indicate independent contextual statements does not provide any additional expressivity to the statements already made, it allows a curator to capture several statements from the same evidence within a paper. An example is when specifying the multiple substrates of an enzyme—the enzyme may act on each of the substrates independently, but not all at the same time; therefore, the substrates can be listed in the extension separated by pipe symbols:

Gene product:	UniProtKB:O14522 (*human PTPRT*)
GO term:	GO:0004725 (*protein tyrosine phosphatase activity*)
Extension:	has_direct_input(UniProtKB:P12830) | has_direct_input(UniProtKB:O60716) (*E-cadherin*|*CTNND1*)

This annotation indicates that the receptor protein tyrosine phosphatase rho (PTPRT) dephosphorylates E-cadherin and CTNND1, but not necessarily both simultaneously. It would be equally correct to create two separate annotations each with a single substrate in the extension.

5 Practical Use of Extended Annotations

There are likely to be many use cases for extended annotations—even some we have not yet envisioned. Users will be able to perform more advanced queries with the available functional data; such as filtering on the subcellular, cellular or anatomical locations in which a gene product performs its roles, or which genes a transcription factor regulates in a specified cell type. Annotation extensions can also help create functional networks through the use of directional relationships such as has_input and has_direct_input, which allow specification of the target of an effector, for example in a signaling pathway or the substrates of a metabolic enzyme activity.

Without contextual detail, bioinformatics analyses of gene products involved in a specified process cannot distinguish, for example, between those gene products that are active only in a particular cell type and those that are inactive or absent from that cell type, therefore creating a bias in the interpretation of the data. With extended annotations any differences in the active components of a process or pathway between various cell or tissue types can be determined.

5.1 Access

Extended annotations are available for download in the current GO annotation files, both in the GAF2.0 format (column 16; http://www.geneontology.org/GO.format.gaf-2_0.shtml) and in the Gene Product Association Data format (GPAD column 11; http://www.geneontology.org/GO.format.gpad.shtml). These files can be accessed from the GOC website (http://geneontology.org/GO.downloads.annotations.shtml) and the GOA website (http://www.ebi.ac.uk/GOA/downloads).

Extended annotations can be accessed on the web via the GO browsers QuickGO ([18]; www.ebi.ac.uk/QuickGO-Beta) and AmiGO 2 ([19]; http://amigo.geneontology.org/amigo/). Both browsers allow users to filter annotation sets based on the contents of the annotation extension. The display of extended annotations may be different depending on the resource (Fig. 1), but the GO annotation files display the plain text extension since this is more compatible for computational analysis (*see* also Chap. 11 [20]). Any questions on how to access or use extended annotations should be directed to the GOC helpdesk (http://geneontology.org/form/contact-go).

5.2 Exercise

The addition of extended annotations to Gene Ontology datasets enables users to perform sophisticated queries. This exercise will demonstrate how to build such a query in the GOC browser

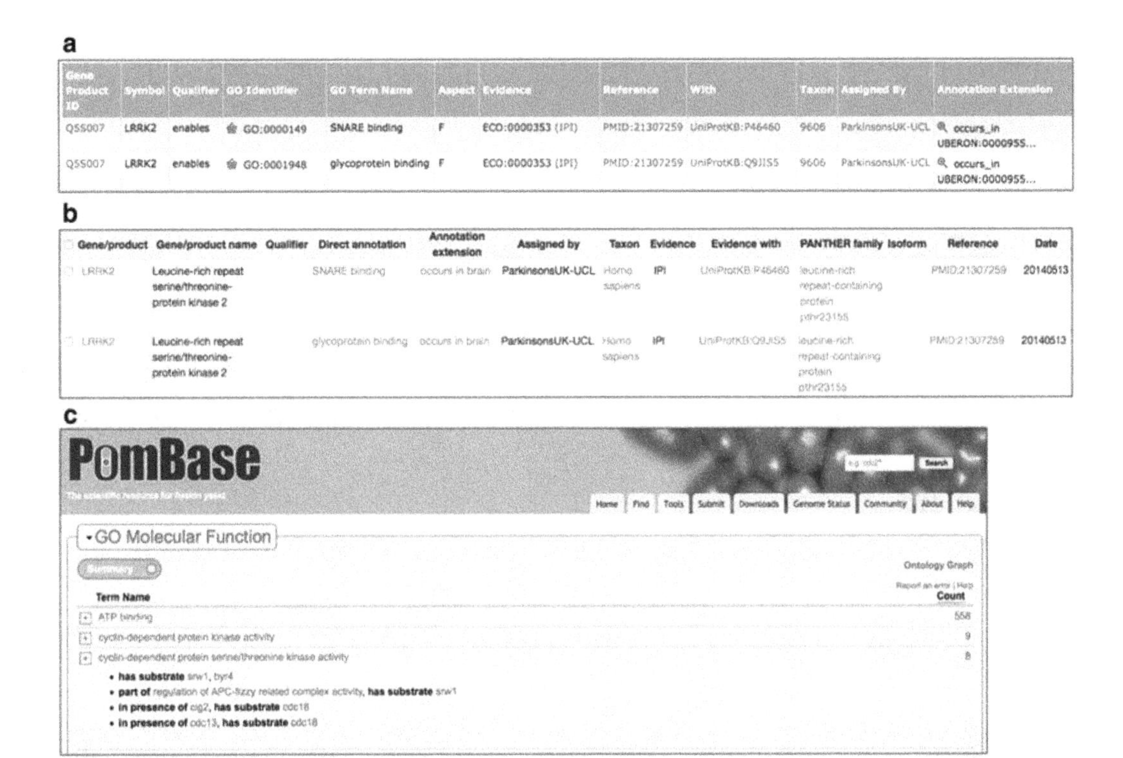

Fig. 1 Display of extended annotations in (**a**) the beta version of the EBI GO browser QuickGO, (**b**) AmiGO 2, and (**c**) PomBase (http://www.pombase.org/)

AmiGO 2, namely, to provide all of the gene products from *S. pombe* that are located in the spindle midzone during mitotic anaphase.

1. Open the AmiGO 2 browser (http://amigo.geneontology.org/amigo/).
2. Click on the Advanced Search button and select "Annotations" from the drop-down list.
3. In the free-text filtering box on the left (Fig. 2a) type in GO:0051233, the GO identifier for the Cellular Component term "spindle midzone".
4. Now open the Taxon menu on the left and click on the "more" button at the bottom. A pop-up menu will open, in the top filter box start typing "pombe"--"Schizosaccharomyces pombe" should be the only option that appears. Click on the + next to the species name to add this to the filter.
5. Now open the Annotation Extension menu on the left and click on the + button next to the term "mitotic anaphase" to add this to the filter.
6. AmiGO 2 will display all of the annotations that use the "mitotic anaphase" term (or one of its child terms) in the annotation extension of a primary annotation to "spindle midzone" (or one of its child terms) (Fig. 2b).

Fig. 2 Finding annotations in AmiGO 2 based on annotation extension data. (**a**) Filters applied in the AmiGO 2 browser: GO:ID (GO:0051233 "spindle midzone"), annotation extension (mitotic anaphase), taxon (Schizosaccharomyces pombe). (**b**) Results of the search using the filters applied in (**a**). Six unique gene products are located to the spindle midzone during mitotic anaphase

6 Summary

The Gene Ontology has proven a vital resource for researchers, enabling them to easily find and use functional data. GO is continually evolving to reflect both accumulating biological knowledge and the computational techniques that researchers need for analysis of a list of gene products. One of the major limitations of using the original simple GO functional annotation has been the lack of contextual information linking together gene products and the roles and pathways they are involved in [4]. Inclusion of this type of data within GO annotations can advance pathway and network analyses substantially, allowing more sophisticated queries and analyses to be performed.

As with all other aspects of GO, annotation extensions continue to evolve—through discussion involving all GOC members and the community—to allow representation and ultimately simple access to a wide variety of contextual data.

References

1. Khatri P, Drăghici S (2005) Ontological analysis of gene expression data: current tools, limitations, and open problems. Bioinformatics 21:3587–3595. doi:10.1093/bioinformatics/bti565
2. Schmidt A, Forne I, Imhof A (2014) Bioinformatic analysis of proteomics data. BMC Syst Biol 8(Suppl 2):S3. doi:10.1186/1752-0509-8-S2-S3
3. Bauer S (2016) Gene-category analysis. In: Dessimoz C, Škunca N (eds) The gene ontology handbook. Methods in molecular biology, vol 1446. Humana Press. Chapter 13
4. Khatri P, Sirota M, Butte AJ (2012) Ten years of pathway analysis: current approaches and outstanding challenges. PLoS Comput Biol 8:e1002375. doi:10.1371/journal.pcbi.1002375
5. Huntley RP, Harris MA, Alam-Faruque Y et al (2014) A method for increasing expressivity of Gene Ontology annotations using a compositional approach. BMC Bioinformatics 15:155. doi:10.1186/1471-2105-15-155
6. Meehan TF, Masci AM, Abdulla A et al (2011) Logical development of the cell ontology. BMC Bioinformatics 12:6. doi:10.1186/1471-2105-12-6
7. Mungall CJ, Torniai C, Gkoutos GV et al (2012) Uberon, an integrative multi-species anatomy ontology. Genome Biol 13:R5. doi:10.1186/gb-2012-13-1-r5
8. Avraham S, Tung C-W, Ilic K et al (2008) The Plant Ontology Database: a community resource for plant structure and developmental stages controlled vocabulary and annotations. Nucleic Acids Res 36:D449–D454. doi:10.1093/nar/gkm908
9. Lee RYN, Sternberg PW (2003) Building a cell and anatomy ontology of Caenorhabditis elegans. Comp Funct Genomics 4:121–126. doi:10.1002/cfg.248
10. The UniProt Consortium (2014) UniProt: a hub for protein information. Nucleic Acids Res 43:D204–D212. doi:10.1093/nar/gku989
11. McDowall MD, Harris MA, Lock A et al (2015) PomBase 2015: updates to the fission yeast database. Nucleic Acids Res 43:D656–D661. doi:10.1093/nar/gku1040
12. Meldal BHM, Forner-Martinez O, Costanzo MC et al (2014) The complex portal—an encyclopaedia of macromolecular complexes. Nucleic Acids Res. doi:10.1093/nar/gku975
13. Hastings J, de Matos P, Dekker A et al (2013) The ChEBI reference database and ontology for biologically relevant chemistry: enhancements for 2013. Nucleic Acids Res 41:D456–D463. doi:10.1093/nar/gks1146
14. Eyers CE, McNeill H, Knebel A et al (2005) The phosphorylation of CapZ-interacting protein (CapZIP) by stress-activated protein kinases triggers its dissociation from CapZ.

Biochem J 389:127–135. doi:10.1042/BJ20050387

15. Iskandar BJ, Rizk E, Meier B et al (2010) Folate regulation of axonal regeneration in the rodent central nervous system through DNA methylation. J Clin Invest 120:1603–1616. doi:10.1172/JCI40000
16. Gloerich M, ten Klooster JP, Vliem MJ et al (2012) Rap2A links intestinal cell polarity to brush border formation. Nat Cell Biol 14:793–801. doi:10.1038/ncb2537
17. Xu N, Papagiannakopoulos T, Pan G et al (2009) MicroRNA-145 regulates OCT4, SOX2, and KLF4 and represses pluripotency in human embryonic stem cells. Cell 137:647–658. doi:10.1016/j.cell.2009.02.038
18. Binns D, Dimmer E, Huntley R et al (2009) QuickGO: a web-based tool for Gene Ontology searching. Bioinformatics 25:3045–3046. doi:10.1093/bioinformatics/btp536
19. The Gene Ontology Consortium (2010) The Gene Ontology in 2010: extensions and refinements. Nucleic Acids Res 38:D331–D335. doi:10.1093/nar/gkp1018
20. Munoz-Torres M, Carbon S (2016) Get GO! retrieving GO data using AmiGO, QuickGO, API, files, and tools. In: Dessimoz C, Škunca N (eds) The gene ontology handbook. Methods in molecular biology, vol 1446. Humana Press. Chapter 11

PERMISSIONS

All chapters in this book were first published by Springer; hereby published with permission under the Creative Commons Attribution License or equivalent. Every chapter published in this book has been scrutinized by our experts. Their significance has been extensively debated. The topics covered herein carry significant findings which will fuel the growth of the discipline. They may even be implemented as practical applications or may be referred to as a beginning point for another development.

The contributors of this book come from diverse backgrounds, making this book a truly international effort. This book will bring forth new frontiers with its revolutionizing research information and detailed analysis of the nascent developments around the world.

We would like to thank all the contributing authors for lending their expertise to make the book truly unique. They have played a crucial role in the development of this book. Without their invaluable contributions this book wouldn't have been possible. They have made vital efforts to compile up to date information on the varied aspects of this subject to make this book a valuable addition to the collection of many professionals and students.

This book was conceptualized with the vision of imparting up-to-date information and advanced data in this field. To ensure the same, a matchless editorial board was set up. Every individual on the board went through rigorous rounds of assessment to prove their worth. After which they invested a large part of their time researching and compiling the most relevant data for our readers.

The editorial board has been involved in producing this book since its inception. They have spent rigorous hours researching and exploring the diverse topics which have resulted in the successful publishing of this book. They have passed on their knowledge of decades through this book. To expedite this challenging task, the publisher supported the team at every step. A small team of assistant editors was also appointed to further simplify the editing procedure and attain best results for the readers.

Apart from the editorial board, the designing team has also invested a significant amount of their time in understanding the subject and creating the most relevant covers. They scrutinized every image to scout for the most suitable representation of the subject and create an appropriate cover for the book.

The publishing team has been an ardent support to the editorial, designing and production team. Their endless efforts to recruit the best for this project, has resulted in the accomplishment of this book. They are a veteran in the field of academics and their pool of knowledge is as vast as their experience in printing. Their expertise and guidance has proved useful at every step. Their uncompromising quality standards have made this book an exceptional effort. Their encouragement from time to time has been an inspiration for everyone.

The publisher and the editorial board hope that this book will prove to be a valuable piece of knowledge for researchers, students, practitioners and scholars across the globe.

LIST OF CONTRIBUTORS

Paul D. Thomas

Pascale Gaudet, James C. Hu and Christophe Dessimoz

Janna Hastings

Patrick Ruch

Sylvain Poux

Domenico Cozzetto and David T. Jones

Gemma L. Holliday, Rebecca Davidson, Eyal Akiva and Patricia C. Babbitt

Iddo Friedberg and Predrag Radivojac

Richard J. Roberts and Martin Steffen

Alex Warwick Vesztrocy

Catia Pesquita

Fran Supek and Nives Škunca

Sebastian Bauer

Monica Munoz-Torres and Seth Carbon

Marcus C. Chibucos, Deborah A. Siegele and Michelle Giglio

Rachael P. Huntley and Ruth C. Lovering

Index

Printed in the USA
CPSIA information can be obtained
at www.ICGtesting.com
JSHW051351091023
49903JS00006B/117